# Technology and International Relations:
# Challenges for the 21st Century

# Technology and International Relations:

# Challenges for the 21st Century

*Dr. Bhaskar Balakrishnan*

*Former Ambassador of India*

**Vij Books India Pvt Ltd**
New Delhi (India)

**Indian Council of World Affairs**
Sapru House, New Delhi

*Published by*

**Vij Books India Pvt Ltd**
(Publishers, Distributors & Importers)
2/19, Ansari Road
Delhi – 110 002
Phones: 91-11-43596460, 91-11-47340674
Fax: 91-11-47340674
e-mail: vijbooks@rediffmail.com
web : www.vijbooks.com

Paperback reprint 2018

ISBN: 978-93-86457-32-5

*To my wife Shobhana, and children*
*Prashant and Nitya.*

# Contents

# Foreword I

This book is a pioneering and very well-researched effort on a subject of great current interest. We must take advantage of the present friendly political attitude towards India in developed countries, by providing the right environment for investment in high technology fields. Similarly we should be ready to help less-developed countries. The Director General of the International Atomic Energy Agency Yukiya Amano said in a lecture in the Bhabha Atomic Research Centre that "India is at the forefront of technological development in the nuclear sector, not least in the area of fast reactors and related fuel cycles... India's remarkable success in the field of peaceful nuclear technology is an inspiration for many developing countries...I also appreciate India's willingness to serve as a mentor for other Asian countries that have recently joined the IAEA."

The range of topics covered in this book is remarkable and the description and the conclusions by him drawn are based on reliable information, backed by references. I would like to mention in particular the important international conventions covering various aspects related to technology, ranging from space, ocean, and chemical weapons to climate change threat, information on any of which is not easy to come by. India's role in the negotiations leading to these conventions and other issues related to science and technology are circumscribed by the fact, as Dr. Balakrishnan says, that "diplomats seem unfamiliar with the science subjects while scientists seem unfamiliar with international ramifications of scientific developments affecting national interests." To close this gap is the main objective of this book. He, however, mentions that

"India's foreign policy in recent times has become more pragmatic and focused on meeting the needs of India's development".

The book gives a concise, and easily understandable, account of the history and recent advances in a variety of technology fields, which have led to changes in society. Dr. Balakrishnan asks the rhetorical question: "Can technological change be controlled and adjusted to maximise benefits and minimize disruptions in society?" We must also remember that technology domination is sought both by countries and by companies through the mechanisms of intellectual property rights and technology control regimes. These issues are addressed by the author in later chapters. I paraphrased a long time back the futurologist Alvin Toffler's famous statement *"Yesterday violence was power, today wealth is power and tomorrow knowledge will be power"* to say that "Today, more than at any time in history, *'Technology is Power'* and this will continue to be so in the foreseeable future". I said this because all the sources of power Toffler mentions have their foundations in technology.

The chapter on Nuclear Technology is very well-written, dealing with all aspects of nuclear power and nuclear weapons. While Dr. Balakrishnan was in the Foreign Service, he spent some time in Vienna, and our embassy there deals both with Austria as well as the International Atomic Energy Agency; so the author has first-hand experience with nuclear matters.

He talks of the importance of chemical industry, and also refers to the green chemistry movement. He describes how biotechnology has grown rapidly with the possibility of international cooperation providing a great opportunity to transform agriculture and health in developing countries. I liked in particular his analysis of topics like the independent intergovernmental organization ICGEB, the Human Genome Project and the Biological Weapons Convention.

India has leadership in many high-technology fields like nuclear, space and IT software. The author describes these as well as the role of social media and cyber security and how the internet can help in digital manufacturing. Readers will also learn about how the UN Commission on the Limits of the Continental Shelf defines the rights of the coastal states to harvest mineral and non-living material in the subsoil of the continental shelf. Such information is not easy to come by. The author has also touched upon other technology fields like nanotechnology.

Talking of international scientific cooperation, Dr. Balakrishnan asks: "Can science bring us together?" The present trend is that global mobility of students and scientists is increasing. Scientific papers with authorship from more than one country are also increasing rapidly. "Collaborative Innovation" was the theme of the World Economic Forum meeting in Davos in 2008. The author correctly quotes, as examples, the wide-ranging international cooperation in the Centre for European Nuclear Research in Geneva, where the Large Hadron Collider saw a few years back the first signature of the Higgs Boson, and in the International Thermonuclear Experimental Reactor (ITER), coming up in Cadarache in France. These are excellent examples and India is a member of both the projects.

The author concludes with his perception that science and technology are likely to play an increasing role in diplomacy and international relations. I completely agree and would add that, as India develops rapidly and moves towards a knowledge economy, its role in global diplomacy would correspondingly increase.

Dr. Balakrishnan has written an excellent and very readable book, readable by both the layman and the professional scientist.

Dr. R. Chidambaram
Principal Scientific Adviser to the Government of India and
Chairman, Scientific Advisory Committee to the Cabinet
March 2017

# Foreword II

This book represents and unusual and remarkable effort to survey recent trends and developments in an exceptionally wide range of science and technology areas and their likely impact on foreign policy in the future. It is true as the author states that diplomats may not understand how science may impact foreign policy, while scientists may not understand the foreign policy implications of scientific development. The public, policy makers, and the business community often reacts at a late stage when there are already disruptions to their lives and business profits due to science and technology. For them also, this book could be interesting reading.

As an example, take the case of genetically engineered crops. Since ages, man has been developing genetically modified crops (GMOs) through natural processes which produce random genetic changes. Now technology such as CRISPR has given us the ability to make highly precise changes in genes, conferring unique and useful attributes to plants and across species. Higher productivity, resistance to drought, disease and pests, and inclusion of useful nutritional agents such as vitamins can be generated in food crops. There is even the possibility of expanding the range of principal food crops far beyond rice, wheat, and maize. The era of synthetic genes is on the horizon. Concerns have arisen among some farmers and civil society groups that this new technology may be used by powerful multinationals to dominate and abuse the market for enhanced crop production, as well as cause disrupt ions in traditional farming and loss in biodiversity. These are valid concerns, but need to be addressed on the basis of scientific evidence and informed discussion. The battle over GMOs is a good example of disruptive technology that has cross border ramifications requiring policy makers to respond.

Climate change is another area where science, technology and multiple disciplines are involved, posing a serious threat to the survival of mankind. If indeed we are all on spaceship Earth, then we need to cooperate or endanger everyone. Delays in responding adequately will only make future generations suffer more. The author has brought out the challenge of climate change in clear terms and stressed the need for urgent action.

Similarly, human health, nanotechnology, information technology, and climate change are areas of rapid progress and are likely to throw up challenges to society in general and in foreign policy. Policy makers could in future be overwhelmed by the sheer pace of technological progress and the multiple level disruptions that this will cause. This will add to the problems of policy making and technology adaptation in developing countries. This book attempts to look at a wide range of science and technology sectors and related issues. The question of how to optimally manage science and technology in the context of foreign policy is also interesting. Can science and technology bring warring nations together, as for example the Israel/Arab states or India/Pakistan? Have we explored these possibilities so far? The author gives some examples of this effort which the author has called techno-diplomacy.

In conclusion I would say that this is a remarkable book well worth reading not only for diplomats and policy makers, but also for the scientific community, business community and the general public.

Dr. M S Swaminathan
Founder Chairman, M S Swaminathan Research Foundation
Ex-Member of Parliament (Rajya Sabha)

# Preface

This book has its origin in a lecture I delivered to students at the Indian Institute of Technology, Roorkee, India on the role of Technology in India's foreign relations. While preparing the lecture, I was struck by the wide range of science and technology issues that had become the subject of international negotiations over the years. It seemed likely that this trend would continue and even accelerate in the future. This raised a number of questions. How were negotiations on science related matters being conducted? Was there enough preparation for such negotiations and were concerns of all stakeholders being taken on board? How prepared were developing countries such as India to handle such negotiations in terms of protecting their long term interests?

It seemed that the interactions between diplomats handling these matters and the scientific community were less than optimal, and largely event or project driven. Diplomats seemed unfamiliar with the science aspects while scientists seemed unfamiliar with the international ramifications of scientific developments affecting national interests. Other sectors such as business and civil society seemed to be actively involved in a narrow sense, when there were immediate challenges to their perceived interests. Advanced countries, particularly the US had evolved fairly elaborate structures and policies for dealing with science and diplomacy in an integrated fashion, and promoted their interests during international negotiations. Given the fact that science and technology would play an expanding role in the future, it seemed appropriate to embark on a project to examine in more detail the implications for diplomacy in general and Indian foreign policy in particular.

This book represents a modest effort at examining a fairly vast and diverse subject, starting from the past and looking towards the future. A number of S & T related sectors have been examined, in varying detail, to draw some conclusions and highlight areas where diplomacy may be required. Also examined is the question of how science and technology related issues could be managed in terms of the diplomatic machinery and various constraints in terms of human and financial resources. The analysis could be relevant for any developing country, not just India. Of course, it is not possible to cover such a vast canvas adequately in one book, as each of the chapters could easily be extended to a full book, and therefore there would be many gaps that could be filled and details elaborated. Given that science and technology is rapidly advancing there could be several recent advances and developments that have not been covered, though I have made every effort to keep it as up to date as possible.

It is my hope that this book would help diplomats to better understand some of the science and technology related issues that have come into the arena of international relations in one way or other. It may also help scientists and technologists to better appreciate the complexities of international relations and the impact that science and technology have on them. For the business community, it may provide motivation for the need for awareness and preparing for the disruptions caused by technology changes as well as for seizing the opportunities afforded by such changes. For the public it may provide a better understanding of how scientific advances can affect their lives and society in general.

I wish to acknowledge the support extended to this project by the Indian Council for World Affairs (ICWA). The cooperation and support extended by the Indian Missions in Moscow and Tel Aviv for a field visit is gratefully acknowledged. I owe a special thanks to my wife, Shobhana and my children Prashant and Nitya, for having believed in me and encouraged and supported me in undertaking this work, as well as for their frank comments and advice.

Dr. Bhaskar Balakrishnan

# Chapter 1

# Concepts, Ideas and Historical Perspectives

*"Scientia Potestas Est – Knowledge is Power"*

*– Francis Bacon, 1597*

The men waited anxiously at the bunker at Alamogordo, New Mexico, on 16th July 1945. Suddenly there was a great flash of light, followed by the thunderous roar of an explosion, and then a blast wave of hurricane strength. A huge multi-coloured mushroom cloud of dust rose over the horizon. Into the mind of J. R. Oppenheimer, the scientific leader of the Manhattan Project (under which the test called Trinity was carried out), flashed a verse from the Indian Bhagavad Gita[1] "If the radiance of a thousand suns were to burst into the sky, that would be like the splendor of the Mighty One..." Seconds later, on seeing the mushroom cloud rising, another verse sprang into his mind "I am become Death, the shatterer of worlds." The age of nuclear weapons technology had dawned. Mankind for the first time had gone far beyond all earlier wars and had developed the capacity to destroy all life on the planet. How would mankind manage this in the uncertain future ahead? Could the system of nation states adapt to this new reality and if so how? Would this make war between nation states unthinkable? Here was a technology that had changed the world beyond all expectations, much before people, societies and nations were prepared for it.

The story of technology is as old as human civilization. From the earliest days, man has constantly tried to enlarge his knowledge, and apply it in diverse ways to meet his needs. The starting point is the quest for knowledge, to understand nature and this is what can be called science (in Latin "scientia" means "knowledge"). Science today means knowledge about nature that can be rationally explained and tested through methods that can be reliably reproduced. The scientific method as understood in modern times is the means of obtaining knowledge about nature through observation, experimentation, logical thinking, and verification of predictions based on the acquired knowledge.

Knowledge about the working of natural things was gathered long before recorded history and had led to the development of complex abstract thinking. This is shown by the construction of complex calendars, techniques for making poisonous plants edible, and buildings such as the pyramids. Ancient India had a sophisticated body of knowledge in its texts, covering mathematics, astronomy, medicine, etc. Ancient China gave to the world four great inventions – printing, paper making, gunpowder, and the compass. However a deeper understanding of why these inventions worked was lacking. What mattered was that it worked in practice.

Much of the processes and thinking that underlay the acquisition of this ancient knowledge have been lost, while details of transmission of this knowledge across ancient societies are sketchy. The Greek philosopher Socrates developed the analytical method of inquiry based on questioning and elimination of hypotheses. Aristotle took this further through the application of logical reasoning and deductive inference, but the link to observation and experimentation was not laid. The modern scientific approach was pioneered by thinkers such as Ibn Al-Haytham (Al Hazem) of Iraq (965-1040 CE) who based his conclusions on reproducible experimentation and analysis of data[2].

Having gained a degree of understanding or grasp of the science underlying some aspect of natural phenomena, the next logical step is to find practical uses for this knowledge. This practical application of science is what we call technology. The word "technology" comes from the Greek word "techne" which means "art, skill, cunning of hand", and 'logia' which means "study of something". Technology has gained prominence after the industrial revolution. It can be defined broadly as "the practical application of knowledge, especially in a particular area, or the capability given by the practical application of knowledge"[3]. This definition is adequate for most discussions.

Throughout history, the search for knowledge and its application through technology have been important determining factors in the progress of human society. In the competition for dominance and control, societies which forged ahead in mastery of basic knowledge and technology were able to succeed, sometimes far beyond expectations. Success could be measured in terms of military superiority over adversaries, or economic advancement through new useful products not available to competing countries. Societies with advances in technology were able to gain advantages over those which did not have technology.

This raises some important questions. What are the conditions under which some societies and individuals within them make technological advances? Does the attitude of the state and ruling elites influence science and technology development? Do resources play a role? Is the level of education in a society relevant? What drives persons to engage in the pursuit of science and technology? Conversely, are there some conditions that militate against advances in science and technology? The analysis of these questions and issues may throw up interesting insights.

Certain conditions are clearly necessary. There must be a minimum level of order and stability in society. Education must be accessible to a substantial number of persons. A constant interchange of ideas and thoughts between societies is important and this requires means of free communication and dissemination of ideas. A certain level of prosperity is required, which enables the generation of surpluses which can be used for research and inquiry. Encouragement and social recognition of thinkers and scientists is an important requisite. Governance should encourage scientific pursuits or at least avoid repressing it. The existence of external and internal challenges to society, including the threat of war, gives a powerful stimulus to creative thinking to meet such challenges.

This paradigm changing, force multiplying and transformative effect of technology has been responsible for major historical changes and relations among societies as well as transformations within societies. Examples are numerous. The discovery of agriculture allowed for the feeding of larger and settled populations, allowing people to be freed from the daily quest for food, and freeing creative energies for art, science and culture. The time available for thinking and innovation increased, and this led to the development of complex societies. The invention of the stirrup in China in the 5th century greatly increased the effectiveness of mounted fighters[4]. The Mongols developed a composite bow which was compact and more powerful, and could be used on horseback giving mobility to a hitherto static long range weapon. Gengis Khan's warriors could shoot targets accurately from a range of 300 metres, and penetrate armour from a distance of 30 paces[5]. Babur the founder of the Mughal Empire in India incorporated Ottoman artillery and tactics into his military which played a key role in his victory in the First Battle of Panipat, 1526. The use of cannons was decisive in ending the role of elephant cavalry in warfare in India. The impact of technology on warfare and military balance was particularly striking, allowing relatively smaller forces to prevail because of superior technology. This phenomenon continues even in modern times, when the first atomic bombs were used to demonstrate to Japan, Germany and the Soviet Union, the enormous power of the new weapon in the hands of the US[6].

Technology has sparked revolutionary changes in society. It can generate new products that can spread rapidly through society. It can increase productivity and work efficiency. Technology changes may be accompanied by material changes as well as changes in management, learning, social interactions, financing, methods of research etc. Technological revolution can change the material conditions of human existence and also reshape culture, society and even human nature. It can play a role of a trigger of a chain of disruptive and unpredictable changes. The consequences of a technological revolution may be positive when it leads to wealth creation, as well as negative, when for example, it causes unemployment or closure of obsolete industries. Technological change is a continuous process that involves innovation at all stages of research, development, diffusion and application.

This raises some interesting issues. Can technological change be controlled and adjusted to maximize benefits and minimize disruptions in societies? Is technological change a spontaneous phenomenon which has its own dynamics? What conditions are required for a society to exploit technological change? How do individuals and organizations adapt to technological changes? What is the role of inter-state cooperation in technological change?

Any new technology is basically a challenge to the existing order based on older technology. Historically governments have not been able to control the disruptive effects of technological change and have been focused on dealing with the effects of such changes. There are very few examples of the state leading the adjustment process to a new technology. The only exception might be in the area of military applications of new technology, where states have led the process of technological adaptation in their armed forces to gain advantages. By and large, the process of technological change has been left largely to market forces which enabled more competitive products and businesses to prevail. In the process other products and businesses had to close down causing job losses and economic hardships. The idea of a state led mechanism to optimize the effects of technological change has not taken root, though states have sought to intervene sometimes to soften the impact of job losses. By the time a new technology has been recognized as disruptive it is usually too late to control the effects of its diffusion. Good examples of this are the ICT revolution and the biotechnology revolution.

Inter–state cooperation in matters of technological change usually begin after a new technology causes disruptions in trade, or threatens the environment, or stability and peace. Such was the case with nuclear weapons which threatened world peace and survival. In the early stages of technology development there is more likely to be competition and secrecy as states strive to gain advantages over others. Once the technology has matured and becomes widespread, the focus shifts to maximizing benefits, while adverse effects tend to be ignored or downplayed. Once the adverse effects are widely understood, interstate cooperation grows to deal with these adverse consequences. However, in recent times, there has been a growth in anti-technology groups who instinctively reject new technology and highlight its adverse consequences. This is the case, for example, with the groups agitating against genetically modified crops, on the grounds of its adverse impact on ecology and traditional farming.

Ancient China witnessed several advances in science and technology including inventions such as the compass, gunpowder, papermaking, and printing which came to Europe some 1000 years later. Gunpowder became the basis of a whole range of weapons, while printing ignited tremendous changes. There were many other engineering innovations in China which led to products such as matches, iron plough, sluice gates, the propeller, etc. Ancient India witnessed considerable advances in irrigation and water management, mathematics and astronomy, medicine and surgery, etc. The ability to innovate and develop technology and apply it contributed to the prosperity of these countries.

The question as to why ancient Asia which pioneered many inventions could not maintain its leadership in technology has puzzled many. While Europe progressed through the industrial revolution and the age of modern science, Asia remained relatively backward as far as science, technology and innovation was concerned. A number of reasons have been advanced such as political instability, prolonged conflicts, attitude of ruling elites, and absence of challenge to the prevailing order. Europe's progress led to its states acquiring military superiority and ultimately to control over much of Asia.

In his path breaking book, "Guns, Germs, and Steel: The fate of human societies", Jarred Diamond[7] argues that gaps in power and technology between human societies originated in environmental differences, which are amplified by various reinforcing factors. The rise of Eurasia according to this theory was due to the availability of more food grain varieties and animal species suitable for domestication over a larger East-West landmass. Europe's geography favored smaller, closer, nation-states, bordered by natural barriers of mountains, rivers, and coastline. Threats posed by immediate neighbours ensured governments that suppressed economic and technological progress soon corrected their mistakes or were weeded out relatively quickly, whilst the region's leading powers changed over time. Other advanced cultures developed in areas whose geography was conducive to large, monolithic, isolated empires, without competitors that might have forced the nation to reverse mistaken policies such as China banning the building of ocean-going ships. Western Europe also benefited from a more temperate climate than Southwestern Asia where intense agriculture ultimately damaged the environment, encouraged desertification, and hurt soil fertility. Competition among European nations encouraged them to innovate and avoid technological stagnation, which led to the dominance of European powers over the last 500 years.

However, even in Europe, the inquisition of the Catholic Church, set up during the 12th – 16th centuries to protect Catholicism against the inroads of other religious ideologies, severely restricted new thinking and ideas. For example, Giordano Bruno, a leading astronomer, was burned at the stake in 1600 after the Roman Inquisition found him guilty of heresy. Galelio Galelei was put on trial for postulating that the Earth moved around the Sun[8]. In medicine, the Catholic Church retarded progress in understanding the functioning of the human body. However the Reformation led to an end to doctrinal shackles on scientific thinking and led to a period of great advances in science and technology. Even in modern times, the dogmatic approach of the Stalinist period in the Soviet Union gave a setback to research in heredity and biology[9].

International relations comprise the variety of relationships among states within the international system and the entire range of global issues, including the roles of governmental and nongovernmental, national and multinational entities. Power is a key factor in the calculus of international relations. It can be described in terms of control over key resources, capabilities, and influence in international affairs. It has been divided up into the concepts of hard power and soft power[10], hard power relating primarily to coercive power, such as the use of force, and soft power commonly covering the persuasive domain, such as economics, diplomacy and influencing people. Technology can lead to a major increase in military capability which can enhance hard power. It can also increase the economic output or knowledge resources of a society and can thus increase soft power. Technology therefore plays a critical role in determining power, both hard and soft.

It is therefore clear that the extent of mastery over basic knowledge and technology are key determinants of a society's military and economic strength, and therefore its ability to participate effectively in the international system. During the Cold War, the western countries sought to restrict export of technology to the Soviet bloc countries, the COCOM[11]. Another was the effort to control the access to technology by developing countries, to leverage economic benefits[12]. However, such leadership in technology cannot remain indefinitely, since knowledge and technology can be acquired by other competitors, albeit with some time delays. Efforts to control or limit the spread of technology are therefore bound to be temporary. Nation states can devote sufficient resources to acquiring technology deemed as vital to their interests. During the Cold War, export restrictions by the West stimulated a drive for self sufficiency by the Soviet Union. The imposition of an embargo on nuclear technology to India stimulated indigenous development of nuclear technology[13]. Similarly, merely acquiring of technology without the capability to derive it from basic knowledge offers only limited advantages, and may even negatively impact genuine indigenous capability. Therefore maintaining leadership in knowledge and technology requires a continuous and sustained effort at building knowledge and innovation.

The quest for knowledge and technology requires not merely material resources. Numerous examples highlight the key role played by human resources, especially of innovative thinkers and researchers. There is a distinction between mastering the "content" or "hardware" of knowledge, and being able to "innovate and apply" or the "software" of knowledge. This phenomenon is found in all disciplines. One can distinguish between a technically perfect musician and a musical genius, a technically well trained athlete and a star performer; and a scientist or engineer who knows the content and one who can also innovate and move beyond limits. Both are important – mastery over content as well as ability to innovate.

The pursuit of technology requires innovation and improvisation, the ability to question conventional assumptions and beliefs, and move ahead into uncharted areas. For example in the early 20 th century, the fundamental conventional assumptions of classical physics were challenged and overthrown, and a whole generation of physicists developed quantum mechanics and relativity. This spirit of challenge and enquiry continued in physics, leading to many major advances. A.P.J. Abdul Kalam has called this process the "igniting of minds", by which one can soar beyond the framework of conventional knowledge and explore new horizons[14].

The structure and attitude of the state plays a key role in the development of science and technology. Throughout history, states which have had enlightened and progressive ruling elites have been able to derive benefits from creative thinkers in various fields, including science and technology. On the other hand when the elites were dogmatic and strictly tied to religious or political doctrines that demanded unquestioning compliance, creativity and innovation suffered. The rise of liberal democratic political thought coincided with the industrial revolution that witnessed a flourishing of science and technology in Europe. Can the free spirit of scientific inquiry flourish under an autocratic and dictatorial state? It is more difficult in such systems for creativity which is essential for scientific development to flourish, although the state may well provide special facilities for this purpose. Restrictions on civil liberties may hamper communications that could affect the speed and extent of technology diffusion and innovation. Diversity and freedom to develop and express new ideas, freedom to travel are critical. Democracies generally have higher technology-induced economic growth than dictatorships, even if they have high quality institutions[15].

## Endnotes

1  Oppenheimer (nicknamed Oppie) had studied Sanskrit and had read the Bhagavad Gita and other Indian Sanskrit holy literature. These verses are in Chapter 11 of the Gita, where Lord Krishna, serving in human form as a chariot driver, reveals and explains his full universal form to the warrior Arjuna.

2  "The 'first true scientist", Jim Al-Khalil,  BBC News.4 January 2009 URL: http://news.bbc.co.uk/2/hi/7810846.stm  Retreived 24-11-2014

3  "Definition of technology", Merriam Webster, URL: http://mw1. merriam-webster.com/dictionary/technology  Retrieved 20-11-2014.

4  Medevial Technology and Social Change, L.T.White, Oxford University Press, p15-16, 1962.

5  Eric, Brownstein. "The Path of the Arrow". http://digitalcollections. sit.edu/cgi/viewcontent.cgi?article=1064&context=isp_collection Retrieved 4 October 2015.

6  The use of the atomic bomb against Japan is a subject of dispute among historians. Some feel that Japan would have surrendered in any case after the Soviet Union entered the war against Japan, and that the atomic bomb was used primarily to demonstrate US power. The Real Reason America Used Nuclear Weapons Against Japan, 12 October 2012,     http://www.globalresearch.ca/the-real-reason-america-used-

nuclear-weapons-against-japan-it-was-not-to-end-the-war-or-save-
lives/5308192 accessed 8-2-2016.

7   Guns, Germs, and Steel: The Fates of Human Societies, Jared Diamond,
E.W.Norton & Co., 1999.

8   Giordano Bruno proposed that the stars were just distant suns surrounded
by their own exoplanets and raised the possibility that these planets could
even foster life of their own. He also insisted that the universe is in fact
infinite and could have no celestial body at its "center". Beginning in
1593, Bruno was tried for heresy by the Roman Inquisition on charges
including denial of several core Catholic doctrines. The Inquisition
found him guilty, and in 1600 he was burned at the stake in Rome. In
1616, the Roman Inquisition found Copernicus's proposition that the
sun is immobile and at the center of the universe and that the Earth
moves around it to be "foolish and absurd in philosophy" and that the
first was "formally heretical" while the second was "at least erroneous in
faith". This assessment led to Copernicus's work to be banned. Galileo
Galilei revised those same theories and was also admonished for his
views on heliocentrism. In 1633, the Roman Inquisition tried Galileo
and found him "vehemently suspected of heresy". Galileo died under
house arrest.

9   Lysenkoism or Lysenko-Michurinism was the centralized political
control exercised over genetics and agriculture by Trofim Lysenko and
his followers. Lysenko was the director of the Soviet Union's Lenin All-
Union Academy of Agricultural Sciences. Lysenkoism began in the late
1920s and formally ended in 1964. Lysenkoism was built on theories
of the heritability of acquired characteristics that Lysenko named
"Michurinism". These theories depart from accepted evolutionary
theory and Mendelian inheritance. Lysenkoism is used metaphorically
to describe the manipulation or distortion of the scientific process as a
way to reach a predetermined conclusion as dictated by an ideological
bias, often related to social or political objectives.

10 Soft Power, Joseph S Nye, Foreign Policy, 80, 153-171, 1990; Soft Power - The means to success in world politics, Joseph S Nye Jr., Perseus Books, 2004.

11 The Consultative Group-Coordinating Committee (COCOM) was established in 1949 by 14 NATO countries plus Japan, in order to impose a collective embargo of strategic exports to Soviet bloc countries by all Western countries.

12 Achieving Nonproliferation Goals: Moving From Denial to Technology Governance, Dr. Elizabeth Turpen, June 2009, Stanley Foundation, http://www.stanleyfoundation.org/publications/pab/turpenpab609.pdf accessed 10-2-2016

13 Controlling International Technology Transfer: Issues, Perspectives, and Policy Implications, Tagi Sagafi-Nejad, Richard W. Moxon, Howard V. Perlmutter, Elsevier, Oct 2013, p 229, 237

14 Ignited Minds- Unleashing the Power Within India, A. P. J. Abdul Kalam, 2002

15 Governance and Knowledge: The Politics of Foreign Investment, Technology and Ideas, Helge Hveem, Carl Henrik Knutsen, Routledge, 2012, p14-18

# Chapter 2

# Technology in the 20th Century Society and Impact on International Relations

*"Knowledge itself ... turns out to be not only the source of the highest-quality power, but also the most important ingredient of force and wealth. Put differently, knowledge has gone from being an adjunct of money power and muscle power, to being their very essence. It is, in fact, the ultimate amplifier. This is the key to the powershift that lies ahead, and it explains why the battle for control of knowledge and the means of communication is heating up all over the world",*

*– Alvin Toffler, Powershift, 1990.*

The immense achievements of technology by 1900 and especially the fertile period after 1867 were greatly surpassed in the twentieth century which witnessed more advances over a wide range of activities than the whole of previously recorded history. The airplane, the rocket and interplanetary probes, electronics, atomic power, antibiotics, insecticides, and a host of new materials have all been invented and developed to create an unparalleled social situation, full of possibilities and dangers, which would have been virtually unimaginable in earlier times[1]. It was a period of unprecedented advances in science and technology, which profoundly changed human civilization. Scientific discoveries, such as the theory of relativity and quantum physics, overturned the foundations of physics. It was a century that started with radio, aircraft, automobiles, and ended with high-speed rail, cruise ships, global commercial air travel and the space shuttle.

Mass media, telecommunications, and information technology especially computers and the Internet made the world's knowledge more widely available. Advancements in medicine improved the welfare of people with global life expectancy increasing from 35 years to 65 years. The impact on society and international relations was tremendous. It is convenient to divide the century into two periods. The first period from 1900-1945 saw two World Wars, while the second from 1945 to 2000 went by without a world war, though there were numerous conflicts and threats to peace.

In the field of military affairs technological advancement transformed warfare, enabling it to reach unprecedented levels of destruction. World War II killed over 60 million people, while nuclear weapons gave mankind the means to destroy itself. The Imperial system was shattered, the era of empires and wars of expansion and colonization ended, resulting in a far more globalized and interdependent world.

During World War I, technology began to play an important role in military affairs. A period of scientific discovery and innovation and the industrial revolution had ushered in a period of prosperity. The development of weapons such as machine guns and artillery led to trench warfare using heavily defended lines. To overcome these, technologies such as poison gas were developed and countered with gas masks. 20th century technology with 19th century warfare resulted in heavy casualties which gave way to more effective integration of technology in warfare with mobile armoured fighting vehicles making their appearance. Aircraft fitted with guns began to be used. At sea, the submarine made its appearance while battleships were more heavily armoured and used longer range guns.

During the interwar years, Germany and the Soviet Union were dissatisfied powers that for different reasons cooperated with each other on military research and development. The Treaty of Versailles had imposed severe restrictions upon Germany constructing vehicles for military purposes, and so throughout the 1920s and 1930s, German arms manufacturers and the Wehrmacht had begun secretly developing tanks. The Soviets offered Weimar Germany facilities deep inside the USSR for building and testing arms and for military training, away from treaty inspectors. In return, they asked for access to German technical developments, and for assistance in creating a Red Army General Staff. The artillery manufacturer Krupp was soon active in the south of the USSR. In the late 1920s, Germany helped Soviet industry begin to modernize, and to assist in the establishment of tank production facilities.

Britain sold hundreds of its best aircraft engines to German firms which used them in a first generation of aircraft, and then improved on them much for use in German aircraft. These new inventions contributed to German success in World War II. Göttingen was the main center of theoretical and mathematical aerodynamics and fluid dynamics research from 1904 to the end of World War II. Germany's advances in internal combustion engine development and aerodynamics and fluid dynamics contributed to the development of jet aircraft and of submarines with improved under-water performance. To cope with shortages of petroleum resources, Germany had developed a process to make synthetic fuel from coal.

The role of technology increased greatly in World War II. Some of the technologies used were developed during the 1920s and 1930s; some were developed during the war, while others were beginning to be developed as the war ended. Major developments occurred in several fields such as Weaponry and delivery platforms, Logistical support systems, Communications and Intelligence, Medicine, and Military Industry. World War II was the first war where military operations widely targeted the research efforts of the enemy. This included the exfiltration of scientists; the sabotage of heavy water production in Norway; and the bombing of rocket facilities at Peenemunde. Military operations were also conducted to obtain intelligence on the enemy's technology; for example, the German radar and for the German V-2 rocket.

World War II marked the first full-scale war where mechanization and mobility played a significant role. Most nations were not equipped for this. Even the German Panzer forces relied heavily on non-motorized support and flank units, though Germany recognized and demonstrated the value of concentrated use of mechanized forces. The British also saw the value in mechanization as a way to enhance an otherwise limited manpower reserve. America sought to create a mechanized army using its strong industrial base that could afford such equipment on a great scale. The most visible vehicles of the war were the tanks, forming the armored spearhead of mechanized warfare. Their impressive firepower and armor made them the premier fighting machine of ground warfare. There were a large number of trucks and lighter vehicles that kept the infantry, artillery, and others moving.

Naval warfare changed dramatically during World War II, with the ascent of the aircraft carrier to the premier vessel of the fleet, and increasingly capable submarines. Submarines were critical in the Pacific Ocean as well as in the Atlantic Ocean. Advances in submarine technology included the snorkel. Japanese defenses against Allied submarines were ineffective. Much of the merchant fleet of the Empire of Japan was sunk. The Kriegsmarine had developed the pocket battleship to get around constraints imposed by the Treaty of Versailles. Innovations included the use of diesel engines, and welded rather than riveted hulls. Important advances in the field of anti-submarine warfare such as technologies for the detection of submarines through sonar became widespread as did shipboard and airborne radar.

A large array of guns, mortars, artillery, bombs, and other devices were developed during the war to meet specific needs that arose, but many traced their early development to prior to World War II. Torpedoes began to use magnetic detonators; compass-directed, programmed and even acoustic guidance systems; and improved propulsion. Fire-control systems continued to develop for ships' guns and came into use for torpedoes and anti-aircraft fire. A wide range of improved small arms, semi automatic and machine guns were also developed.

The Allies of World War II cooperated extensively in the development and application of new and existing technologies to support military operations and intelligence gathering. Examples of cooperation include the Sherman Firefly and the American-led Manhattan Project. British technological developments in fields such as radar, jet propulsion and also the early British research into the atomic bomb were made available to the US.

The Manhattan Project, a massive research and development effort to quickly develop an atomic bomb, or nuclear fission warhead was perhaps the most profound military development of the war, and had a great impact on the scientific community, among other things creating a network of national laboratories in the United States. In 1942 the United Kingdom dispatched around 20 British scientists and technical staff to America, along with their work. The scientists formed the British contribution to the Manhattan Project, where their work on uranium enrichment was instrumental in jump-starting the project. In August 1945, two atomic bombs were employed against the Empire of Japan, causing unprecedented destruction and civilian casualties.

There was also a German nuclear energy project, including talk of an atomic weapon. This failed for a variety of reasons, most notably German Anti-Semitism. Half of continental theoretical physicists who did much of their early study and research in Germany, were either Jewish or, in married to a Jew. Some left Germany for political reasons. The few remaining scientists left lacked the high morale that characterized the Los Alamos work. This illustrates the negative impact of political control over free thought on scientific innovation and research.

The strategic importance of the bomb, and its even more powerful fusion-based successors, did not become fully apparent until the United States lost its monopoly on the weapon in the post-war era. The Soviet Union developed and tested their first nuclear weapon in 1949. Nuclear competition between the two superpowers played a large part in the development of the Cold War. The strategic implications of such a massively destructive weapon continue in the 21st century. The collaboration between the British and the Americans led to the 1958 US-UK Mutual Defence Agreement between the two nations, whereby American nuclear weapons technology was adapted for British use.

In the post World war II era, the advent of nuclear weapons, and the rivalry between two power blocs posed a challenge to the international system. While the basic science behind nuclear weapons was fairly simple, in practice the construction of nuclear weapons required formidable technical efforts in production of fissile material and weapons design expertise. The nuclear weapons arms race resulted in numerous nuclear tests and the rise of USSR, UK, France, and China as nuclear weapons states. The spectre of a nuclear war with unimaginable consequences led to intense public pressure on governments to prevent this threat to human survival.

For these reasons, the area of nuclear arms control was the first where technology impacted international relations in a major way after World War II. This led to various arms control negotiations such as the NPT, the Nuclear Test Ban treaties, covering nuclear technology, as well as Treaties on Chemical and Biological weapons, etc, that covered the broad area of weapons of mass destruction. These negotiations involved science and technology related issues, and required technical knowledge, especially in the verification and compliance areas.

Another key area of research was aerospace, which had already led to development of rockets and high performance aircraft, used to deliver weapons. Human capability expanded to being able to place platforms far above the earth in space, on various missions, including going to the moon. This led to concerns over military use of space and international efforts to prevent armed conflict in space. Spacecraft launch vehicles could as well deliver nuclear weapons across the globe. Space based assets could be used for intelligence as well as useful remote sensing purposes. The area of peaceful uses of outer space had to be enlarged as much as possible. Another concern over spread of missile technology was sought to be curbed by evolving a plurilateral missile technology control regime among the technology possessing countries. As human capability expands to cover possible manned missions to the moon and planets, international cooperation will be seen as mutually beneficial and will require negotiations.

In the 1970s international debate focused on the North-South divide, with the developing countries under the umbrella of the G-77 demanding fundamental changes in the international order. One key area of discussion was the question of access to technology on fair and equitable terms. Technology was seen as the key to economic development and higher value addition. Multinational corporations which had the technology would only grant access to it on their terms, which often included restrictive business practices that went against the competition laws of their own countries. Efforts were undertaken to evolve an international code of conduct for transfer of technology but these did not yield success, as many western governments were reluctant to impose transnational guidelines on private enterprises.

The discovery of the structure of genetic material such as DNA as a common underlying structure of all life forms on earth opened up huge possibilities. For the first time human beings had the power to alter the process of evolution and produce new life forms. This development had transformative impact on life sciences, human and animal health, agriculture, environment, and led to the new field of biotechnology and genetic engineering. This raised numerous issues at the international level, such as regulation of transboundary movement of genetically modified organisms, research into potentially dangerous bioagents, etc. Ethical issues related to human genetic information, embryonic stem cell research, altering human the genome, creation of new life forms, came into prominence. Negotiators had to grapple with issues that required a deep knowledge of biology as well as international repercussions. Developing countries were keen to exploit these advances, and international mechanisms were sought to be set up to facilitate the transfer of biotechnology.

In some areas such as fundamental research into the structure of matter, the requirement of funding, equipment and personnel becomes very large, beyond the capacity of individual countries. This has led to international collaboration in large scientific projects which lead to benefits to the participating countries. Examples are CERN, the ITER, the International Space Station, and the Human Genome Project. Such projects require international negotiations and an understanding of the technical issues involved. The list of such projects is likely to grow in future, as mankind delves deeper into the mysteries of nature.

The rapid growth of information and communications technology (ICT) in the 20 st century has led to vast benefits for human society and has transformed it. Access to knowledge has become easier, while access to large numbers of people through ICT has become possible. ICT has become integrated into many areas of human society, from governance, to enterprise functioning and infrastructure management. This has exposed vulnerabilities to cyber attacks and frauds. ICT has the potential for military use. The international community is faced with the problem of evolving regimes to deal with these aspects while enabling the beneficial use of ICT.

As the industrialization process that started in the West spread to the developing world, a consequence was the further increase in emission of gases that caused damage to the ozone layer and greenhouse gases such as carbon dioxide, which led to global warming. While much of the emissions came from the industrialized countries, the problem of global warming required a global consensus. Tackling these problems required technical knowledge of the scientific aspects as well as integrating them into the international negotiating frameworks. The Montreal Protocol, one of the success stories, was devised to protect the ozone layer. The UN Framework Convention on Climate change (UNFCC) and its consequent instruments such as the Kyoto Protocol, represented global efforts to negotiate a solution to the climate change problem, but did not yield a definitive solution.

Has spending on S & T kept pace with the rapid pace of development? There is general agreement that a reasonable level of expenditure on research and development should be in the range of 2 percent of GDP, including funding from public and sources. Of course, the amount of expenditure is only one factor, it also matters how the resources are used. Analysis of the spending trends[2] throws up some interesting facts. China raised its R & D expenditure the most going from 0.57 percent of GDP in 1996 to 2.05 percent in 2014. Similar figures (percent of GDP) for some other economies - Korea 4.29 in 2014 (from 2.24 in 1996), Israel 4.09 in 2014 (2.6 in 1996), Japan 3.58 in 2014 (2.77 in 1996), India 0.82 in 2013, (0.63 in 1996), USA   2.73 in 2013 (2.44 in 1996), Russia 1.19 in 2014 (0.97 in 1996), and World average 2.12 in 2013, (1.99 in 1996). In terms of absolute R&D spending (in PPP billion US$ in 2014), the leading countries are USA (485) China (344), Japan (163), Germany (103), Korea (64), India (62) France (58), Russia (54). China is closing the gap with the US while the other countries are relatively far behind.

The rapid rise of China in S & T to challenge the US is indeed a remarkable feature of the 21 st century. What will be the consequences of this rise? As the 21 st century unfolds, it has become clear that the acceleration of science and technology development will have an increasing impact on international systems, challenging human ingenuity in finding solution.

# Endnotes

1   History of Technology, R.A.Buchanan, Encyclopedia Brittannica, https://www.britannica.com/technology/history-of-technology/The-20th-century , accessed 13-4-2017

2   2016 Global R&D Funding Forecast, Industrial Research Institute (IRI), https://www.iriweb.org/sites/default/files/2016GlobalR%26DFu ndingForecast_2.pdf , accessed 13-4-2017

# Chapter 3

# India's Technological and Foreign Policy Evolution

*"By 2030 India will be among the top three countries in science and technology and will be among the most attractive destinations for the best talent in the world. The wheels we set in motion today will achieve this goal."*

*- Narendra Modi, Prime Minister of India,*
*at the Indian Science Congress 2017.*

## Science in Ancient India

Ancient Indian scholars are said to have developed geometric theorems before Pythagoras and were using advanced methods of determining the number of mathematical combinations by the second century B.C. By the fifth century A.D., Indian mathematicians were using ten numerals and by the seventh century were treating zero as a number. The conceptualization of geometric shapes, fractions, and the ability to express the number ten to the twelfth power, algebraic formulas, and astronomy had even more ancient origins in Vedic literature, some of which was compiled as early as 1500 B.C[1].

Technological discoveries made in India relate to pharmacology, surgery, medicine, artificial colors and glazes, metallurgy, chemistry, the decimal system, geometry, astronomy, and language and linguistics. These discoveries led to practical applications in brick and pottery making, metal casting, distillation, surveying, town planning, hydraulics, the development of a lunar calendar, etc.

Archaeological evidence and written accounts from other cultures with which India has had contact support the evidence of Indian scientific and technological developments. The technology of textile production, hydraulic engineering, water-powered devices, medicine, and other innovations, as well as mathematics and other theoretical sciences, continued to develop and be influenced by techniques brought in from the Muslim world by the Mughals after the fifteenth century

Technology applications include thousands of water tanks for irrigation in South India by the eighteenth and nineteenth centuries forming a closely integrated network supplying water throughout the region. In metallurgy numerous small but sophisticated furnaces were built for producing iron and steel. By the late eighteenth century, production capability is estimated to have reached 200,000 tons per year. India was the world's leading producer and exporter of textiles before 1800.

Indians had built a network of sophisticated, large-scale astronomical observatories, the Jantar Mantars in the early eighteenth century. These complexes were used to determine the seasons, phases of the moon and sun, and locations of stars and planets from points in Delhi, Mathura, Jaipur, Varanasi, and Ujjain. The Jantar Mantars were designed and built by Sawai Jai Singh II, the ruler of Amber, between 1725 and 1734.

## Indian Science and Colonialism

The arrival of the British in India in the early seventeenth century along with the Portuguese, Dutch, and French led eventually to new scientific developments that added to the indigenous achievements of the previous centuries. Colonization turned India into a source of raw materials for the factories of England and France and left only low-technology production to local entrepreneurs. However, the British education system brought in Science education which expanded to meet the need for trained indigenous people. What new technologies were implemented were imported rather than developed indigenously.

Western education and techniques of scientific inquiry were added to the already established Indian base. The major result of these developments was the establishment of a large educational infrastructure in India which continued to produce generations of top scientists. The British education system, aimed at producing able civil and administrative services candidates, exposed a number of Indians to foreign institutions. Sir Jagadish Chandra Bose (1858–1937), Prafulla Chandra Ray (1861-1944), Satyendra Nath Bose (1894–1974), Meghnad Saha (1893–1956), P. C. Mahalanobis (1893–1972), Sir C. V. Raman (1888–1970), Subrahmanyan Chandrasekhar (1910–1995), Homi Bhabha (1909–1966), Srinivasa Ramanujan (1887–1920), Vikram Sarabhai (1919–1971), Har Gobind Khorana (1922–2011), and Harish Chandra (1923–1983) were among the notable scholars of this period

One of the most famous scientists of the pre and post-independence era was Indian-trained Chandrasekhara Venkata (C.V.) Raman, an ardent nationalist, prolific researcher, and writer of scientific treatises on the molecular scattering of light and other subjects of quantum mechanics. In 1930 Raman was awarded the Nobel Prize in physics for his 1928 discovery of the Raman Effect, on inelastic scattering of light which gave useful information on molecular structure and led to an entire branch of molecular spectroscopy. He was a director of the Indian Institute of Science and founded the Indian Academy of Sciences in 1934 and the Raman Research Institute in 1948.

Another leading scientist was Homi Jehangir Bhabha, an eminent physicist internationally recognized for his contributions to the fields of positron theory, cosmic rays, and muon physics at the University of Cambridge in Britain. In 1945, with financial assistance from the Sir Dorabji Tata Trust, Bhabha established the Tata Institute of Fundamental Research in Bombay.

Other eminent pre-independence scientists include Sir Jagadish Chandra (J.C.) Bose, a Cambridge-educated Bengali physicist who discovered the application of electromagnetic waves to wireless telegraphy in 1895 and then went on to a second notable career in biophysical research. Meghnad Saha was trained in India, Britain, and Germany and became an internationally recognized nuclear physicist whose work provided new insights into ionized gases. In the late 1930s, Saha began promoting the importance of science to national economic modernization, a concept fully embraced by Nehru and several generations of government planners.

Satyendranath Bose was trained in India, and his research discoveries gave him international fame and an opportunity for advanced studies in France and Germany and collaboration with Albert Einstein. The Bose-Einstein Statistics, used in quantum physics and Boson particles are named after him. Prafulla Chandra Ray, another Bengali, earned a doctorate in inorganic chemistry from the University of Edinburgh in 1887 and went on to a devoted career of teaching and research. His work was instrumental in establishing the chemical industry in Bengal in the early twentieth century.

Srinivasa Ramanujan (1887-1920) was a remarkable Indian mathematician with almost no formal training in pure mathematics, who made extraordinary contributions to mathematical analysis, number theory, infinite series, and continued fractions. Ramanujan initially developed his own mathematical research in isolation; it was quickly recognized by Indian mathematicians. When his skills became apparent to the wider mathematical community centred in Europe at the time, he began a famous partnership with the English mathematician G. H. Hardy. During his short life, Ramanujan independently compiled nearly 3900 results (mostly identities and equations). Nearly all his claims have now been proven correct, although some were already known. He stated results that were both original and highly unconventional, such as the Ramanujan prime and the Ramanujan theta function, and these have inspired a vast amount of further research.

India had several internationally renowned scientists in the Indian Diaspora in the US such as Subrahmanyan Chandrasekhar (1910-1995), an Indian American astrophysicist born in Lahore, Punjab. Chandrasekhar was awarded, along with William A. Fowler, the 1983 Nobel Prize for Physics, with Chandrasekhar cited for his mathematical theory of the physical processes of importance to the structure and evolution of the stars. This work led to the currently accepted theory on the later evolutionary stages of massive stars, including black holes. The Chandrasekhar limit on stellar mass is named after him. Hargobind Khorana, (1922 – 2011) was an Indian-American biochemist who shared the 1968 Nobel Prize for Physiology or Medicine for research that helped to show how the order of nucleotides in nucleic acids, which carry the genetic code of the cell, control the cell's synthesis of proteins.

The above illustrates the capabilities of Indian scientists and thinkers, despite the lack of institutional structures to support science and technology. Many of India's scientists did their work in relative isolation from the western scientific community. In some case, they benefited from western support and collaboration. In other cases they did their productive work in foreign institutions. The lack of an institutional basis and a community of scientists and technologists hampered the realization of India's full potential.

## Science in independent India

After independence, India's first Prime Minister Jawaharlal Nehru aimed "to convert India's economy into that of a modern state and to fit her into the nuclear age and do it quickly." Nehru understood that India had not been at the forefront of the Industrial Revolution, and hence made an effort to promote higher education, and science and technology in India. Nehru's high priority to science and technology was reflected in his tasking Dr. Homi J Bhabha to put in place a long term strategic plan for nuclear science and technology. With remarkable foresight, Nehru also promoted science and technology institutions, such as the first IIT in Kharagpur in 1951, the Indian Space Research programme in the 1960s, and the CSIR. These institutions led to a growth in India's science and technological capacity. However, bureaucratic management and failure to attract the best of young scientific talent remained a problem.

This thrust in science and technology continued during after Nehru, including the Pokhran I series of nuclear explosions in 1974. India became host to one of the two centres of the International Centres for Genetic Engineering and Biotechnology (ICGEB). Strong political support was extended to science and technology, including information technology and telecommunications, leading to India emerging as the world's leading software services provider.

## Policy Evolution

Science and Technology policy of the Government evolved through various policy statements. The Science Policy Resolution of 1958 was followed by a Technology Policy Statement of 1984. The Science and Technology Policy Statement of 2003, outlined several key objectives in the wake of economic reforms. These included promotion of scientific research in universities and other institutions, and creation of employment opportunities in the S&T sectors, greater participation of women, and greater autonomy and freedom of functioning for academic institutions. Stronger mechanisms were envisaged for technology development evaluation, absorption and upgradation from concept to utilization. The IPR regime would be revised for speedy and effective domestic commercialization of such inventions. Innovation was recognized as a key ingredient. Spending on R & D was targeted at 2 percent of GDP, but the actual level attained was only around 1 percent. In 2013, the Government launched a Science, Technology and Innovation (STI) Policy[2], focused on accelerating the pace of discovery and delivery of science-led solutions for serving the aspirational goals of India for faster, sustainable and inclusive growth. It envisages creation of a new STI ecosystem, which finds solutions to societal problems and facilitates the entire innovation chain from knowledge to wealth creation, while at the same time attracting best students to this area, ensuring a premier position for India in the scientific world

In recent years this thrust in science and technology has been maintained and stepped up despite changes in government. The recent initiatives taken by the government such as "skill India", "make in India", "start up India", ICT enabling of important government functions, "ease of doing business in India", improvements in infrastructure, together with the stronger outreach to Indian diaspora are examples of the effort to build a strong S & T ecosystem. This should produce rich dividends in the future by way of greater manufacturing of high technology products. According to Prime Minister Narendra Modi[3], the Technology Vision 2035 document[4] released in 2016 is now developing into a detailed roadmap for twelve key technology sectors. NITI Aayog is evolving a holistic science and technology vision for India, where technology needs to span a wide range; from advanced technologies to rural development needs.

The driving force behind India's science and technology initially came from government initiatives such as those in atomic energy, space, and biotechnology. A large network of government research laboratories were set up under the administration of the Ministry of Science and Technology, the Council of Scientific and Industrial Research, and the Ministry of Defence. The private sector was relatively less active, engaged more in the education sector in engineering and medical colleges. The problem remains of managing government run scientific establishments in a way that yields optimum results in terms of attracting scientific talent and producing the best output. In later years, the private sector was the driver in areas such as information technology. In the area of scientific and technical manpower, the IITs proved to be a successful model. But many of the IIT graduates, faced with limited science and technology opportunities in India, sought and obtained work abroad, or moved into other sectors of the economy, constituting a "brain drain".

Science and Technology research and development does not take place in isolation. It is embedded into the overall S & T ecosystem which is a subset of the national economic ecosystem. A mature S & T ecosystem contains not only scientific institutional and human resources in various entities such as research centres and academic institutions both public and private, but also entities that help commercialize and support the results of S & T, such as financing sources for research, entrepreneurs, business support agencies, regulatory and government agencies, and venture capital agencies. The full benefits from S & T research require that the various components of this ecosystem function effectively. A large part of research conducted in India does not yet achieve its full exploitation due to gaps in this ecosystem.

## Foreign Policy Evolution

The principles of India's foreign relations were initially formulated by Nehru. These included non-alignment with either of the two blocs in the Cold War, peaceful coexistence and constructive collaboration with all countries irrespective of their internal governance systems, and securing largest possible space for India's indigenous economic, scientific and technological development considered vital for removal of poverty. This policy served India well, enabling it to build cooperation with both sides in the Cold War, while largely avoiding entanglement in the East-West conflicts that resulted during this period. However, India's foreign policy continues to face challenges in dealing with border disputes and armed conflicts with two of its neighbours, China and Pakistan.

The Cold War has faded away, and in its wake there are multiple poles of power – the US, China, EU, Russia, and Japan with changing power equations among them. The global situation is marked by instability, rise of non-state actors with transnational capability, states with failing governance, global threats such as climate change, etc. At the same time there is growing competition and lack of collaboration among nations, which poses formidable challenges in the area of foreign and security policy. Recent trends are of growing sentiment against globalization and towards isolationism, even in the US and Europe, which could further aggravate the situation.

India's foreign policy in recent times has become more pragmatic and focused on enabling and meeting the needs of India's development, and securing the maximum in terms of strategic autonomy. With India's economy growing at a faster rate, India has been more actively engaged[5] with major global issues, and with major economies. India's rising economic and political influence on the global stage has led to a much heavier and active agenda in terms of foreign affairs, posing challenges of capacity for India's foreign policy institutions. At the same time the old challenges remain, of securing peace and stability in the immediate neighbourhood, managing relations with its neighbours especially Pakistan and China, and combating the growing menace of terrorism, extremism, and climate change. Science and Technology issues, which lie at the core of many global challenges, will continue to play an important role in India's foreign policy in the years ahead.

## Endnotes

1  James Heitzman and Robert L. Worden, editors. India: A Country Study. Washington: GPO for the Library of Congress, 1995.

2  Science, Technology and Innovation Policy, 2013, Department of Science and Technology, Government of India, http://www.dst.gov.in/sites/default/files/STI%20Policy%202013-English.pdf , accessed 13-4-2017

3  PM's Address at the Inauguration of the 104th Session of the Indian Science Congress, Tirupati, 3 Jan 2017, Press Information Bureau, Government of India, http://pib.nic.in/newsite/PrintRelease.aspx?relid=156086 , accessed 13-4-2017

4   Technology Vision 2035, Technology Information, Forecasting and Assessment Council(TIFAC), January 2016, http://planning.kar.nic.in/docs/SDG/Technology%20Vision%202035.pdf , accessed 13-4-2017

5   Modi doctrine: A new foreign policy for India, Gautam Mukherjee, The Pioneer, 18 Aug 2016, http://www.dailypioneer.com/columnists/edit/modi-doctrine-a-new-foreign-policy-for-india.html , accessed 13-4-2017

# Chapter 4

# Nuclear Technology and India's Foreign Policy- the curse of Prometheus

*"The conflict that exists today is no more than an old-style struggle for power, once again presented to mankind in semireligious trappings. The difference is that, this time, the development of atomic power has imbued the struggle with a ghostly character; for both parties know and admit that, should the quarrel deteriorate into actual war, mankind is doomed."*

— *Albert Einstein, Apr 1955*

The area of nuclear technology probably contains the most significant challenges in India's foreign relations and indeed to human survival. Towards the end of World War II, in Aug 1945, the US detonated nuclear weapons over the cities of Hiroshima ("Little Boy")[1] and Nagasaki ("Fat Man")[2] causing over 200,000 deaths. Japan surrendered soon thereafter to the allies on 15 August 1945. The perception among the proponents of using the nuclear weapons was that it would accelerate Japan's surrender and prevent a large number of casualties during the planned invasion of Japan. Half the deaths occurred immediately within one day, while the other half was over 4 months due to burns, injuries and radiation exposure. Almost all the deaths were of civilians. The experience of Hiroshima and Nagasaki is important in that it is the only occasion that nuclear weapons were used against large cities, a probable outcome in the event of an outbreak of nuclear war. The discovery of nuclear weapons proved to be the proverbial curse of Prometheus for mankind, generations being condemned to live continuously under the threat of annihilation.

An arms race soon broke out involving nuclear weapons among the allied powers, led by the US and the USSR[3]. Numerous nuclear weapons tests conducted in the Pacific, North Africa, and Siberia by the US, UK, France, and the Soviet Union during the period 1949-62. The U.S. Castle Bravo test in 1954 was a new form of hydrogen bomb, with a yield of 15 Megatons—over twice what was predicted. It generated a large amount of radioactive nuclear fallout, more than had been anticipated, and a change in the weather pattern caused the fallout to be spread in a direction which had not been cleared in advance. The fallout plume spread high levels of radiation for over a hundred miles, contaminating a number of populated islands in nearby atoll formations. Though they were soon evacuated, many of the islands' inhabitants suffered from radiation burns and later from other effects such as increased cancer rate and birth defects. Similar problems—unpredictably large yields, changing weather patterns, unexpected fallout contamination of populations and the food supply—occurred during other atmospheric nuclear weapons tests by other countries as well. Injurious effects on human health from the absorption of radioisotopes such as Strontium-90, Cesium-137, etc were soon noticed. Particularly insidious was the effect of these radioisotopes on the very young and fetuses. The exposure to radiation from these isotopes was found to cause increases in cancer and birth defects. These effects took several years to appear.

The concerns over human health effects of radioactive fallout from atmospheric tests soon led to a global clamour for an end to such tests, even if carried out in remote areas. The fallout from any atmospheric test gets carried across the earth by prevailing winds. At the same time, there was a strong demand for the complete abolition of nuclear weapons, articulated forcefully but pragmatically by Prime Minister Jawaharlal Nehru in April 1954, becoming the first world leader to do so[4]. Global pressure led to negotiations for a partial test ban treaty, under which parties committed themselves not to carry out any atmospheric tests. However, by then, the main nuclear powers, the US, USSR, UK, and France had already carried out a large number of nuclear weapons tests, including the atmospheric tests of super bombs "Castle Bravo" in 1954, and the "Tsar Bomba" by the USSR in 1962[5]. These had already released considerable radioactive fallout into the atmosphere. The Partial Test Ban Treaty in 1963[6] limited signatories including the US, UK, and USSR to underground testing. Not all countries stopped atmospheric testing, but because the United States and the Soviet Union were responsible for roughly 86% of all nuclear tests, their compliance cut the overall level substantially. France continued atmospheric testing until 1974, and China until 1980.

A total of over 2,000 nuclear explosions were detonated worldwide between 1945 and 1996. 25 % or over 500 bombs were exploded in the atmosphere: over 200 by the United States, over 200 by the Soviet Union, about 20 by Britain, about 50 by France and over 20 by China. About 75% of the explosions were underground. During the Cold War (1945–1989), the US and USSR conducted over 800 and nearly 500 underground nuclear tests.

Around 20 nuclear tests were conducted by the United States and the Soviet Union at high altitudes or lower outer space between 1958 and 1962. The main aim of these explosions, detonated at heights between 40 and 540 kilometres, was to determine the feasibility of nuclear weapons as anti-ballistic missile defense or anti-satellite weapons. The main destructive effects were through intense electromagnetic pulses (EMP), radiation and charged particles that could destroy electrical and electronic systems. The largest such test, the 1.4 megaton U.S. Starfish Prime test in 1962, damaged and destroyed several of the satellites in orbit at the time and led to widespread power outages on the ground. High-altitude or outer space nuclear testing was subsequently banned by the 1963 Partial Test Ban Treaty as well as by the 1967 Outer Space Treaty.

## Test Ban Treaties

Following the entry into force of the Partial Test Ban Treaty in 1963, nuclear testing continued to be conducted underground by the US, USSR, UK, while France and China conducted nuclear tests both in the atmosphere and underground. In July 1974, the US and USSR signed the Threshold Test Ban Treaty (or TTBT), establishing a nuclear "threshold," by prohibiting nuclear tests of devices having a yield exceeding 150 kilotons. The restraint imposed by the Treaty reduced the explosive force of new nuclear warheads and bombs which could otherwise be tested for weapons systems. Despite the Cold War, the US and USSR could agree on the Antarctic Treaty which entered into force in 1961. This Treaty obligates Parties to refrain from engaging in "measures of a military nature, including testing of any type of weapons," including nuclear explosions, in Antarctica and prohibits the disposal there of radioactive waste material.

## Comprehensive Test Ban Treaty (CTBT)

The PTBT was seen by the opponents of nuclear weapons as an important step towards stopping all nuclear testing and the further development and spread of nuclear weapons. But contrary to expectations, after the PTBT entered into force, the Nuclear Weapons States (NWS) continued their testing of nuclear weapons underground. Progress to move ahead to a comprehensive test ban treaty was blocked due to cold war rivalry which led the NWS to continue testing and nuclear weapons development. After the end of the Cold War in 1991, the Parties to the PTBT held an amendment conference that year to discuss a proposal to convert the Treaty into an instrument banning all nuclear-weapon tests. With strong support from the UN General Assembly, negotiations for a comprehensive test-ban treaty started in 1993. Over the next three years, the Conference on Disarmament in Geneva worked on a draft treaty with two annexures, but failed to reach consensus. India, for its part, stated that it could not go along with a consensus on the draft text and its transmittal to the United Nations General Assembly due to its strong misgivings about the provision on entry into force of the Treaty being conditional on 44 named States (including India) ratifying it[7], which is considered unprecedented in multilateral practice and running contrary to customary international law, and the failure of the Treaty to include a commitment by the nuclear-weapon States to eliminate nuclear weapons within a time-bound framework. Finally the draft treaty was sent to the UN General Assembly which adopted it in 1996 by an overwhelming majority[8]. The Treaty sets up a 51 member Executive Council, and an Organization (the CTBTO) and a Technical Secretariat headed by a Director General, and an International Monitoring System for verification.

By the end of 2015, 164 of the 183 signatory states had ratified the Treaty. Of the 44 states of Annex 2, three (India, Pakistan, and North Korea) have not signed the Treaty, while five States have signed but not yet ratified it (China, Egypt, Iran, Israel, US). Of these, India, Pakistan, and Israel have not signed the NPT and have developed nuclear weapons, while North Korea has withdrawn from the NPT and tested nuclear weapons. Iran remains a party of the NPT but insists on the right to exploit nuclear technology for peaceful purposes, including production of fissile material, which could also be used for nuclear weapons. In the US, the Senate rejected the ratification of the Treaty in 1999, on the grounds that it could compromise national security, and US efforts to modernize its nuclear weapons. Opponents have also rejected moves to get it considered again by the Senate, and called for declaring that the US has no intention to ratify it, thus freeing it of any obligations as a signatory. The CTBT thus has not yet entered into force.

However, its supporters have established a Preparatory Commission for the Comprehensive Nuclear Test-Ban Treaty Organization (CTBTO) as an intergovernmental body in 1996 with its headquarters in Vienna, Austria. The objective of the organization is to achieve the object and purpose of the Treaty, to ensure the implementation of its provisions, including those for international verification of compliance with the Treaty, and to provide a forum for consultation and cooperation among Member States. The Commission prepares for the entry-into-force of the Treaty and carries out the necessary preparations for the effective implementation of the Treaty, including the establishment of a global verification regime. The Preparatory Commission consists of a plenary body composed of all States signatories to the Treaty and a Provisional Technical Secretariat. The PC-CTBTO has been seen as an interim arrangement until the CTBT enters into force. It has grown into a substantial organization with an annual budget of USD 130 million (2015) and a staff of over 260 from over 70 countries. Its International Monitoring System has over 330 installations and uses seismic, hydroacoustic, infrasound and radionuclide data for verification. Conferences on Facilitating the Entry into Force of the Treaty have held in 1999, 2003 and 2007 in Vienna, and 2001, 2005 2009, 2011, and 2013 in New York to maintain pressure to get the Treaty ratified by all states. The UN General Assembly has also been regularly adopting resolutions to this end.

Supporters of the CTBT argue that it has been a success in that while 400 tests were conducted in each of the five decades preceding the adoption of the CTBT in 1996, in the 19 years since then, only a handful have been carried out. In the past decade, only North Korea (DPRK) has tested. However, the NWS continue to hold strategic nuclear arsenals far in excess of their minimum credible deterrence needs. The development of nuclear weapons using methods such as subcritical tests, critical mass experiments, hydronuclear tests, and computer simulations could continue. Critics of the CTBT would argue that the advanced NWS continue to develop nuclear weapons using newer technical means, having obtained all the requisite data through earlier tests, while using the CTBT to prevent other states from acquiring nuclear weapons capability, and maintaining their nuclear weapons dominance.

India's position in not signing the CTBT has led to intense international pressure on it. India's response has been to observe a unilateral moratorium since 1998; it remains an advocate of comprehensive nuclear disarmament; and as stated by Prime Minister Narendra Modi, "will continue to contribute to the strengthening of the global non-proliferation efforts." The Indian nuclear tests of May 1998 have been the subject of much discussion as to whether further tests may be required for strategic reasons. The Shakti I boosted fission device was a test of a two stage fission- fusion nuclear weapon, and some reports indicate that the second stage was not a full success. In comparison, the advanced NWS have developed a three stage fission-fusion-boosted thermonuclear weapon as their standard strategic weapon. This may indicate that India may need to test more devices to reach this level of development. Therefore the non-entry into force of the CTBT enables this option to be kept open, while using the US failure to ratify it as a cover.

## Denuclearization of Outer Space

In 1967, the Outer Space Treaty was concluded, prohibiting States Parties from placing in orbit around the earth any objects carrying nuclear weapons or other weapons of mass destruction[9]. In addition to prohibiting nuclear testing in space, the Outer Space Treaty also prohibits Parties from engaging in military maneuvers on celestial bodies, conducting nuclear tests on celestial bodies, installing weapons systems or constructing military bases on celestial bodies. This treaty effectively made outer space, beyond 100 km from the earth's surface, free of nuclear weapons.

## Denuclearization of Sea Bed

The Seabed Treaty was finalized in 1971 as a multilateral agreement between the United States, Soviet Union (now Russia), United Kingdom, and 91 other countries banning the emplacement of nuclear weapons or "weapons of mass destruction" on the ocean floor beyond a 12-mile coastal zone[10]. It allows signatories to observe all seabed "activities" of any other signatory beyond the 12-mile zone to ensure compliance. However this treaty does not restrict the use of submarines and ships as carriers of nuclear weapons, an important aspect of strategic deterrence. Only the sea bed is denuclearized, while nuclear weapons states are free to send their ships and submarines through the seas with nuclear arsenals.

## Nuclear Non-Proliferation Treaty (NPT)

The nuclear weapons states US, USSR, Britain and France obviously preferred a regime in which such possession of nuclear weapons would be limited to them. The UK had benefited from a joint development effort with the US giving it access to nuclear weapons technology. China derived support from the USSR in this field, at least until the 1960s. France went ahead with developing its own independent nuclear arsenal called the "force de frappe". The intensifying Cold War rivalry reached its most dangerous manifestation in the Cuban Missile crisis of October 1962 which brought the world to the brink of nuclear war. China's nuclear tests in 1964 in the Lop Nor testing grounds in Xinjiang province further accelerated the move for the Nuclear Non-Proliferation Treaty (NPT).

As early as 1959, the United Nations General Assembly adopted a resolution calling for nuclear weapon States (NWS) to refrain from transferring nuclear weapons to non-nuclear weapon States (NNWS). Many observers believed that due to the strategic superiority granted by the atomic bomb, the proliferation of nuclear weapons was inevitable. In 1965 the US proposed the Treaty on the Non-Proliferation of Nuclear Weapons (NPT), while certain states argued that a comprehensive nuclear test ban would cover all related aspects of nuclear nonproliferation. There were disagreements over collective security arrangements; in particular, how the NPT would affect US controlled nuclear weapons deployed in NATO countries. Eventually, the US and USSR recognized that a Treaty on nuclear non-proliferation was in their common interest. The US, U.K. USSR and 58 other countries signed the NPT on 1 July 1968[11]. The Treaty defined nuclear weapon States (NWS) as those countries that tested nuclear weapons before 1967 and all others as non-nuclear weapon States (NNWS). There are three pillars of the NPT: nuclear nonproliferation, nuclear disarmament and the peaceful use of nuclear energy. The Treaty prohibits NWS from transferring nuclear weapons, other nuclear explosives or nuclear weapon technology to NNWS. Likewise, NNWS are obligated to refrain from acquiring nuclear weapons or other nuclear explosive devices. Each NNWS undertakes to accept safeguards agreement with the IAEA. At the same time, NNWS have an inalienable right to nuclear energy for peaceful purposes.

On the key issue of nuclear disarmament, Article VI of the treaty, as well as the preamble, emphasizes disarmament obligations of the NWS. Article VI of the NPT obligates States signatories to "pursue negotiations in good faith on effective measures relating to cessation of the nuclear arms race at an early date and to nuclear disarmament..." The preamble of the NPT recalls the determination expressed by the Parties to the PTBT to "seek to achieve the discontinuance of all test explosions of nuclear weapons for all time and to continue negotiations to this end," foreshadowing the emergence of the Comprehensive Test ban Treaty (CTBT) later on. The nuclear test ban was seen by the NNWS as an important test of whether NWS would live up to their end of the bargain and disarm their nuclear weapons. However, the NWS continued nuclear testing underground until they had reached a point where sufficient test data had accumulated to make weapons design independent of nuclear testing and closing off testing by any potential new entrants to the NWS category could be to their advantage. The NPT does not contain any specific restrictions prohibiting the sharing of nuclear test data and nuclear weapons design information. The NPT laid the foundation of the international nuclear non-proliferation regime. However it has no binding and verifiable means to ensure that NWS would abide by their obligations on disarmament. Therefore, although observers regard the low number of countries that possess nuclear weapons as a great success, NWS continued testing and stockpiling nuclear weapons for decades after the Treaty entered into force. To date, there are still an estimated 27,000 nuclear weapons in the world. Moreover there has been a qualitative improvement in the types of weapons and their diversity of potential use, including tactical nuclear weapons.

This unequal and unbalanced treaty legitimized the possession of nuclear weapons in the hands of five states, arbitrarily defined as those states that had tested a nuclear weapon by 1967. It placed numerous binding restrictions and controls on access to and application of nuclear technology by non nuclear weapons states, coupled with only a mild non-binding obligation on the nuclear weapons states to reduce nuclear arsenals. India, along with several other countries rejected this unequal treaty. They considered the NPT as serving the geostrategic interests of only the five Nuclear Weapons States.

Countries such as North Korea, Iran, and Libya perceiving threats to their security including from the US, a NWS, have sought to keep their nuclear weapons options open while being a party to the NPT. This has brought them into disputes over compliance with the NPT, compounded by mutual distrust and suspicion over their assertion of the right to pursue peaceful uses of nuclear technology, including dual use technology related to fissile material productions, such as uranium enrichment, plutonium separation, and heavy water reactors using natural uranium.

In 1991, following the end of the apartheid regime, South Africa signed the Nuclear Non-Proliferation Treaty. In 1993, it admitted that the country had developed a limited nuclear weapon capability. These weapons were subsequently dismantled before South Africa acceded to the NPT and opened itself up to IAEA inspection. In 1994, the IAEA completed its work and declared that the country had fully dismantled its nuclear weapons program.

Libya had signed and ratified the Nuclear Non-Proliferation Treaty in 1968 and was subject to IAEA nuclear safeguards inspections, but undertook a secret nuclear weapons development program in violation of its NPT obligations, using material and technology provided by the A.Q. Khan Proliferation network, including actual nuclear weapons designs allegedly originating in China. In 2003, Libya reached agreement on eliminating all its WMD programs, and permitted U.S. and British teams (as well as IAEA inspectors) into the country to assist this process and verify its completion. The nuclear weapons designs, gas centrifuges for uranium enrichment, and other equipment were removed from Libya by the United States during 2004. Libya's return to compliance with safeguards and Article II of the NPT was welcomed. In 2011 the Libyan government was overthrown as a result of a military intervention by NATO forces acting under the auspices of United Nations Security Council. Some analysts noted that NATO's intervention in Libya shortly after the nation agreed to nuclear and chemical weapons disarmament would make other countries such as North Korea more reluctant to give up nuclear programs due to the risk of being weakened as a result.

It is clear from the examples above that a major impetus for developing nuclear weapons capability is the sense of threat and insecurity of a country. This was the case with South Africa, Israel, Iran, and North Korea. Indeed, in the case of the NWS, it was the US becoming the first NWS and its conflict with the USSR and its allies that led to the USSR and then China becoming NWS. The UK was a strategic partner with the US and shared nuclear weapons capability. In the case of France, it was President De Gaulle's fierce determination to play an independent global role that drove it to become a NWS. In the case of India, the 1971 threat of US intervention in the Bangladesh war and the perceived nuclear threat from China that drove it to develop nuclear weapons[12]. Pakistan, having lost the 1971 Bangladesh war and a major part of its territory, felt the need to develop nuclear weapons to counter India's conventional weapons superiority and Prime Minister Bhutto declared that the nations would "eat grass if necessary to develop nuclear weapons"[13]. Thus the underlying disease is insecurity while the nuclear weapons capability development is a symptom of that disease. Unless the insecurity component is addressed, limiting the spread of nuclear weapons and indeed other WMDs will require much greater efforts and possibly risk conflict.

## Regionalization of the NPT approach - Nuclear Weapons Free Zones

A regional approach to nonproliferation flowing from the NPT was the concept of nuclear weapons free zones (NWFZ). The first such effort resulted in the Treaty of Tlatelolco, 1967, covering Latin America and the Caribbean. The Parties to the Treaty committed themselves not to manufacture, acquire, test or possess nuclear weapons. There are two additional protocols to the treaty which binds those overseas countries with territories in the region (the United States, the United Kingdom, France, and the Netherlands) to the terms of the treaty, and secondly, requires the worlds declared nuclear weapons states to refrain from undermining in any way the nuclear-free status of the region. This has been signed and ratified by the USA, the UK, France, China, and Russia. Protocol II in Article 3 stipulates that the parties "undertake not to use or threaten to use nuclear weapons against the Contracting Parties of the Treaty for the Prohibition of Nuclear Weapons in Latin America". This provides a significant security guarantee. The Treaty entered into force in 1968, with most of the all the Latin American countries (except Cuba), Jamaica and Trinidad and Tobago signing it. By 1994 all the Caribbean states had also signed and ratified it. Cuba, which had security concerns vis a vis the US, a NWS, signed the treaty only in 1995 and ratified it in 2002. Argentina which had a significant nuclear programme, had an armed conflict with the UK in 1982 over the Falkland Islands, and ratified the Treaty only in 1994, 26 years after signing it. Thus security issues in the case of Cuba and Argentina played a role in delaying their ratification.

The NWFZ concept was sought to be extended to other areas following the model of the Tlatelolco Treaty. The Treaty of Rarotonga, or the South Pacific Nuclear Free Zone Treaty, 1985, formalizes a Nuclear-Weapon-Free Zone in the South Pacific. The treaty bans the use, testing, and possession of nuclear weapons within the borders of the zone. It was signed by the South Pacific nations of Australia, the Cook Islands, Fiji, Kiribati, Nauru, New Zealand, Niue, Papua New Guinea, the Solomon Islands, Tonga, Tuvalu, Vanuatu and Western Samoa in 1985 and has since been ratified by all of those states. The Federated States of Micronesia, Marshall Islands, and Palau are not party to the treaties but are eligible to become parties should they decide to join the treaty in the future. There are three protocols to the treaty, which have been signed by the five declared nuclear states, with the exception of Protocol 1 for China and Russia who have no territory in the Zone. These require (1) no manufacture, stationing or testing in their territories within the Zone (2) no use against the Parties to the Treaty or against territories where Protocol 1 is in force and (3) no testing within the Zone. In 1996 France and the United Kingdom signed and ratified the three protocols. The United States signed them the same year but has not ratified them. China signed and ratified protocols 2 and 3 in 1987. Russia has also ratified protocols 2 and 3 with reservations. The Treaty defines the geographical area to which it applies[14]. The security guarantee contained in Protocol 2 is a significant feature of the Treaty.

The Southeast Asian Nuclear-Weapon-Free Zone Treaty (SEANWFZ) or the Bangkok Treaty of 1995 is a nuclear weapons moratorium treaty between 10 Southeast Asian member-states under the auspices of the ASEAN: Brunei Darussalam, Cambodia, Indonesia, Laos, Malaysia, Myanmar, Philippines, Singapore, Thailand, and Viet Nam. It entered into force in 1997 and obliges its members not to develop, manufacture or otherwise acquire, possess or have control over nuclear weapons. The Zone is the area comprising the territories of the states and their respective continental shelves and Exclusive Economic Zones (EEZ); "Territory" means the land territory, internal waters, territorial sea, archipelagic waters, the seabed and the sub-soil thereof and the airspace above them. The treaty includes a protocol under which the five nuclear-weapon states undertake to respect the Treaty and do not contribute to a violation of it by State parties. None of the nuclear-weapon states have signed this protocol. Article 2 of the Protocol stipulates that each Party "undertakes not to use or threaten to use nuclear weapons against any State Party to the Treaty. It further undertakes not to use or threaten to use nuclear weapons within the Southeast Asia Nuclear Weapon-Free Zone." This is a stronger requirement that the previous two Treaties.

The Treaty of Pelindaba, was concluded in 1996, making Africa a nuclear weapons free zone. It took 32 years to reach this agreement, after the OAU first called for the denuclearization of Africa in 1964. The main reason was the acquisition of nuclear weapons capability by the apartheid regime of South Africa which saw this as a security requirement. It was only after the regime changed in 1992 and South Africa dismantled its nuclear weapons capability and signed the NPT, that the Pelindaba Treaty could be finalized. It has currently been ratified by 39 of the 50 signatory states. Except for the US, all NWS states have ratified Protocol I and II of this Treaty[15], while Protocol III which applies to Spain and France, has not been ratified by either State. The Treaty gives the IAEA an explicit role and establishes an African Commission on Nuclear Energy with a role for settlement of disputes related to the Treaty.

In 1999, the UN adopted a set of guiding principles regarding Nuclear-weapon-free zones which stated that - (1) they should be established on the basis of arrangements freely arrived at among the States of the region concerned; (2) The initiative to establish them should emanate exclusively from States within the region concerned and be pursued by all States of that region; (3) The nuclear-weapon States should be consulted during the negotiations in order to facilitate their signature to and ratification of the relevant protocol(s) to the treaty, through which they undertake legally binding commitments to the status of the zone and not to use or threaten to use nuclear weapons against States parties to the treaty; and (4) A nuclear-weapon-free zone should not prevent the use of nuclear science and technology for peaceful purposes.

In 2006, the five Central Asian states parties to the NPT, were able to finalize a treaty creating a nuclear weapons free zone in Central Asia. The starting point had been the Almaty Declaration of 1997, and progress could be achieved once some of the former Soviet Republics (Kazakhstan) gave up their nuclear weapons. As in the other treaties, there is a Protocol which contains security assurances from the five nuclear weapons states, which have all signed the protocol, with only the US not yet having ratified it.

Efforts to evolve a nuclear weapons free zone in the Middle East and in South Asia have not been successful. The reasons are the existence of security dilemmas in both regions. Israel perceives an existential threat from its Arab neighbours, and has had armed conflicts with them in 1948, 1966 and in 1973. This has driven it to acquire nuclear weapons capability, though undeclared, as a deterrent against aggression, and it has not signed the NPT. Israel is believed to have at least 80 nuclear weapons and stocks of fissile material for another 200 weapons, as well as strategic missiles with ranges up to 7800 kilometres. Israel has refused to negotiate for a nuclear weapons free zone in the absence of a comprehensive peace agreement in the Middle East. On the other hand since 1975, the Arab countries and Iran[16] have been regularly supporting UN General Assembly resolutions calling on all states to join the NPT and to support the creation of a nuclear weapons free zone in the Middle East. Israel remains fiercely vigilant about any attempts by Arab states and Iran to develop nuclear capability, and seeks to minimize it.

In South Asia, the situation is complicated by the existence of two non-NPT signatories India and Pakistan, which have nuclear weapons and have carried out nuclear tests[17]. India is opposed to a nuclear weapons free zone in South Asia as it considers its security being threatened by China, a nuclear weapons state under the NPT, with which it has a border dispute and has had an armed conflict in 1962. Pakistan feels threatened by India's nuclear and conventional weapons forces and has unresolved disputes including over Jammu and Kashmir. The two countries have had armed conflicts in 1948,

1965, and 1971 and an undeclared major conflict in 1999. Therefore the existence security dilemmas and unresolved disputes between China/India and India/Pakistan prevents progress towards a nuclear weapons free zone treaty in South Asia.

There has been some discussion of a nuclear weapons free zone in Central and Eastern Europe, as well as the whole of Europe. Austria and Switzerland had commissioned a study in 2010 on this subject and determined that it could be feasible and would improve security[18]. No recent concrete steps have been taken to discuss the establishment of a NWFZ and 20 years after the End of the Cold War, Europe remains the continent most affected by nuclear weapons. Nuclear weapons from four of the five Nuclear Weapons States (France, United Kingdom, Russian Federation and the United States) are deployed on the European continent. Governments of the region seem unwilling to invest political capital in such a project at the present time. In 1996, Belarus and Ukraine called for a nuclear weapons free zone in Central and Eastern Europe, but this did not get support from countries such as Poland. The security system in Europe has relied heavily on nuclear deterrence and balance between Russia and its allies on the one hand, and the NATO countries with three Nuclear weapons states members. Hence progress towards a nuclear weapons free zone will have to await the evolution of an alternative security system.

## Fissile Material Cut Off Treaty

It had for long been recognized that fissile materials (highly enriched Uranium-235 and Plutonium 239) were the key ingredients of nuclear weapons. Therefore a ban on production of such material could be a way of preventing the growth of nuclear arsenals and could strengthen the NPT regime. The first serious move came in the form of a UN General Assembly Resolution 48/75 of December 1993, which recommended the negotiation of a "non-discriminatory, multilateral, and internationally and effectively verifiable treaty banning the production of fissile material for nuclear weapons or other nuclear explosive devices". This treaty, known as a Fissile Material Cut-off Treaty (FMCT) has been the subject of negotiations in the 65 member Committee on Disarmament (CD) since then. Informal discussions have taken place in the CD on the treaty's purpose; definitions and scope; the production of fissile materials for non-explosive purposes and the role of the International Atomic Energy Agency (IAEA); transparency and stockpiles of fissile materials; compliance and verification; and other provisions including settlements of disputes, entry into force, ratifications, depositaries, duration, and conditions for withdrawal. Nevertheless, the CD was unable to reach consensus and establish a committee to begin formal negotiations on an FMCT.

A major obstacle to launching negotiations has been the issue of existing stocks. While some states, including the United States, United Kingdom, and Japan, favour a treaty which only limits future production of fissile materials, other states, such as those belonging to the Non-Aligned Movement, believe that the treaty should also address fissile materials already produced and stockpiled. This would require nuclear weapon states to irreversibly downblend existing stocks of weapons-grade fissile materials, ensuring they could never be used for weapons purposes again.

Another contentious element is the scope of any potential fissile materials treaty. Although most experts agree that an FMCT would most certainly ban the production of plutonium and highly-enriched uranium, the inclusion of elements such as tritium, for example, which is used to amplify the explosive power of nuclear weapons remains an unsettled question. Since tritium has a radioactive half-life of only 12 years, the inclusion of it in an FMCT would, mean that states would be unable to replace the decaying tritium in existing weapons.

In recent years, Russia, China and the US have modified their positions and are now in favour of moving ahead with the FMCT[19]. In 2004, the United States announced that it opposed the inclusion of a verification mechanism in the treaty on the grounds that the treaty could not be effectively verified. In April 2009, U.S. President Barack Obama reversed the U.S. position on verification and proposed to negotiate "a new treaty that verifiably ends the production of fissile materials intended for use in state nuclear weapons." In May 2009, the CD agreed to establish an FMCT negotiating committee. However, Pakistan has repeatedly blocked the CD from implementing its agreed program of work, despite severe pressure from the major nuclear powers. Pakistan justified its actions arguing that "a proposed fissile material cutoff treaty would target Pakistan specifically.

The United States, United Kingdom, France, and Russia have all declared that they have stopped producing fissile material for nuclear weapons[20]. It is widely believed that China has also stopped producing fissile material for nuclear weapons, ceasing production of highly enriched uranium (HEU) in 1987, and plutonium in 1991. According to the International Panel on Fissile Materials (IPFM), the global stockpile of Highly Enriched Uranium (HEU) in 2011 consisted of roughly 1,500 tons, which would be enough material to create 60,000 simple, first generation nuclear weapons. Roughly 98% of the HEU stock is owned by nuclear weapon states, and Russia and the United States have the largest stocks. India and Pakistan are believed to have ongoing production operations for HEU, although the total global stock continues to decrease, largely because of the efforts of the United States and Russia to down-blend HEU considered to be in excess of military needs. IPFM estimates the global stockpile of separated plutonium at 485 tons, of which, roughly half was produced for use in weapons. The other half was produced for civilian uses. About 98% of plutonium is held by states with nuclear weapons, and the remaining 2% is mostly held by Japan, which has over 10 tons of plutonium. Though the five NWS no longer produce weapons-grade plutonium, production continues in India and Pakistan. Israel still operates its plutonium-producing Dimona reactor, but it is believed to be aimed at producing tritium, rather than plutonium.

There are further complications arising from the use of reprocessing of spent fuel which extracts fissile Plutonium to be fed back as fuel into power reactors. Also fast breeder reactors are under development that can convert Thorium and Uranium into fissile material which can be extracted as reactor fuel. Such fuel cycles are difficult to reconcile with the objectives of the FMCT, as they produce fissile material. The use of reprocessing and fast breeder reactors is considered to be a useful way of getting the most energy production from Uranium and Thorium.

## Nuclear disarmament

There have been calls especially from statesmen in non nuclear weapons states to work towards a complete ban on nuclear weapons. Part of the reason was to prevent nuclear blackmail and threat of use of nuclear weapons against non nuclear weapons states. Another was the realization that a nuclear war even if initially limited could go out of control and could result in destruction of the planet. These efforts have been strongly resisted by the Nuclear Weapons States, who see nuclear weapons as part of their strategic doctrine. They have sought to channelize disarmament initiatives largely along lines that would prevent acquisition of nuclear weapons or capability for them by other states. Hence the advocacy of NPT, its universalization, and promotion of nuclear weapons free zones, etc were given prominence in international agendas. Even the PTBT and CTBT were finalized after the NWS had done most of their nuclear testing and could refine nuclear weapons through virtual testing and computer simulations. Initiatives that could require reductions or dismantling of nuclear arsenals, or could prevent their use were opposed. These included ideas such as no first use, declaration of nuclear weapons as a crime against humanity, etc. After the end of the Cold war, realizing that a new phase in international relations had set in, the nuclear weapons states were more inclined to take measures to reduce nuclear arsenals, which by then were becoming an economic burden, without much additional strategic security. A number of initiatives were undertaken to reduce nuclear arsenals, mainly in the hands of the US and Russia.

US President Obama articulated the idea of a zero option for nuclear weapons. While this idea in principle is what mankind needs, the road to this goal remains elusive and mired in strategic rivalries.

Recently the UN General Assembly adopted a resolution to convene negotiations in 2017[21] on a "legally binding instrument to prohibit nuclear weapons, leading towards their total elimination". Predictably the "nuclear establishment" of the nuclear weapons states and their allies opposed or abstained on this resolution. Civil society groups and public opinion strongly support this resolution. On 23 December 2016, the General Assembly adopted the resolution by a large majority, with 113 UN member states voting in favour, 35 voting against and 13 abstaining. Support was strongest among the nations of Africa, Latin America, the Caribbean, Southeast Asia and the Pacific[22]. A cross-regional group comprising Austria, Brazil, Ireland, Mexico, Nigeria and South Africa[23] initiated the resolution and are likely to lead next year's negotiations. The United States had opposed funding request for the negotiations on the treaty, to be held at UN headquarters in New York, but under intense pressure from supporters of nuclear disarmament, it eventually withdrew its objection. Negotiations on a treaty to ban nuclear weapons under international law began in New York, from 27-31 March 2017, with 132 countries participating and providing inputs, to be followed by a second session from 15 June to 7 July 2017. The president of the conference will produce a draft text which will be subject of final negotiations and adoption[24] of the treaty.

India abstained on the above resolution causing disappointment to many who expected India to vote in favour, given its track record in pushing for universal and complete nuclear disarmament. The vote was puzzling in that if nuclear weapons were to be eliminated completely, India's security vis a vis China and Pakistan as well as other nuclear weapons states would be improved. The explanation of vote suggests that India's hesitancy in supporting this initiative was more on technical rather than ideological grounds. It is hoped that

India will participate actively in the future negotiations towards a nuclear weapons ban treaty.

On 7th July 2017, 122 UN member states voted for a legally binding historic Treaty to ban nuclear weapons. But 69 member states including 8 nuclear weapons states and NATO members refused to participate in the vote on this treaty, while Netherlands voted against and Singapore abstained. Common sense would indicate that the world would be far safer for everyone without nuclear weapons. But the countries that possess nuclear weapons and spend enormous amounts of money on them and associated delivery systems seem to think otherwise despite public pressure to abolish nuclear weapons.

## No First Use (NFU) Policy and Strategic Stability

Pending the agreement on a total ban on nuclear weapons, can something be done to reduce the risk of nuclear war and pursuit of nuclear arsenals? No First Use (NFU) policies have been proposed in this context. No first use (NFU) refers to a pledge or a policy by a nuclear power not to use nuclear weapons as a means of warfare unless first attacked by an adversary using nuclear weapons. This policy seeks to build confidence and raise the threshold for use of nuclear weapons. China declared its NFU policy in 1964, and has since maintained this policy. India declared its policy of no first use of nuclear weapons in 2003. However, NATO has repeatedly rejected calls for adopting NFU policy, arguing that pre-emptive nuclear strike is a key option, in order to have a credible deterrent against the conventional weapon superiority enjoyed by the Soviet Army in the Eurasian land mass.

In 1993, Russia dropped a pledge against first use of nuclear weapons made in 1982. The present Russian military doctrine is similar to that of the US and states that Russia reserves the right to use nuclear weapons in response to the use of nuclear and other types of weapons of mass destruction against it or its allies, and also in case of aggression against Russia with the use of conventional weapons when the very existence of the state is threatened. Earlier, the Soviet

Union had made a no-first-use pledge, later followed by a Russian bilateral agreement with China, but the commitment to a bilateral Russia-China NFU was later relinquished following changes in Russia's nuclear doctrine.

The United States has refused to adopt a no-first-use policy, saying that it "reserves the right to use" nuclear weapons first in the case of conflict. The 2010 Nuclear Posture review reduces the role of U.S. nuclear weapons, stating that, "The fundamental role of U.S. nuclear weapons, which will continue as long as nuclear weapons exist, is to deter nuclear attack on the United States, our allies, and partners." The U.S. doctrine also includes the following assurance to other states: "The United States will not use or threaten to use nuclear weapons against non-nuclear weapons states that are party to the NPT and in compliance with their nuclear non-proliferation obligations. A vigorous debate continues in the US over the NFU policy and its costs and benefits. The UK and France also have policies similar to the US. Some of the US's allies such as Japan and South Korea also oppose a NFU policy by the US as they see it as detrimental to their security which relies on a "nuclear umbrella". Israel does not officially confirm or deny having nuclear weapons, but is widely believed to be in possession of them as well as long range delivery systems. Israel has said that it would not be the first country in the Middle East to formally introduce nuclear weapons into the region.

Pakistan refuses to adopt a "no-first-use" doctrine, indicating that it would launch nuclear weapons even if the other side did not use such weapons first. Pakistan's asymmetric nuclear posture has significant influence on India's ability to retaliate, as shown in 2001 and 2008 crises, when non-state actors carried out deadly terrorist attacks on India, only to be met with a relatively subdued response from India. A military spokesperson stated that Pakistan's threat of nuclear first-use deterred India from seriously considering conventional military strikes. However in 2016 India carried out a military strike in Pakistan controlled Kashmir in response to terrorist operations from that area while Pakistan denied that any strike had taken place. The 1999 Kargil conflict did not see the use of nuclear weapons though it is reported that US President Clinton intervened[25] to press both sides not to use nuclear weapons.

The presence of large nuclear arsenals with India and Pakistan and the high potential for disputes to flare up into armed conflict escalating into a nuclear exchange has aroused considerable concern over the global consequences. While the disputes between the two countries are complicated and may seem intractable, there is good reason to avoid a nuclear exchange that would be extremely destructive. For this reason, after the 1998 tests, Prime Minister Vajpayee[26], in the official paper entitled Evolution of India's Nuclear Policy, expressed India's "readiness to discuss an NFU agreement [with Pakistan] as also with other countries, bilaterally, or in a collective forum." This proposal was reiterated again in August 1998 when India announced that it "will not be the first to use nuclear weapons." In January 2003, the Cabinet Committee on Security reviewed the progress in operationalising India's nuclear doctrine and re-emphasized a posture of NFU thereby underscoring that India would only use nuclear weapons for retaliation. However, Pakistan has consistently refrained from mutually reciprocating India's proposal and retains a first-use posture.

Bilateral NFU agreements with China and Pakistan could contribute greatly to peace and strategic stability in the region, reduce the risk of nuclear conflict, while increasing security for all parties. However, the military-intelligence complex in Pakistan may have to give up its policy of using terrorist groups against India in order for such an initiative to work. Additional complications may arise if indeed Pakistan is holding a stock of dedicated nuclear weapons for other states such as Saudi Arabia. In the case of China, given the commitments made by both countries to a NFU policy, a bilateral NFU could be feasible and prove useful. Similarly a bilateral NFU between India and Russia would also be feasible and could contribute to peace and stability. There is merit in pursuing and promoting a

NFU agreement among all nuclear weapons states, including Israel and North Korea, and possible future nuclear weapons states, as the world would be safer.

## India's Nuclear Technology Development

India's nuclear explosion in May 1974 ("Smiling Buddha"), declared as being for peaceful purposes, used Plutonium separated from a Canadian supplied reactor. This test coupled with India's firm refusal to signing the NPT led to severe restrictions on India's access to nuclear technology, materials and equipment. The Nuclear Suppliers Group was set up to enforce a technology denial regime. This led to a massive indigenous effort under Bhabha Atomic Research Centre (BARC) and the Department of Atomic Energy (DAE) to develop India's strategic and civil nuclear programmes. India's foreign policy in this sphere had to counter the efforts in the international community to isolate and strangle India's fledgling nuclear programme. The main thrust of Indian policy was – to continue to reject the NPT as an unequal and unbalanced treaty, to call for the total abolition of nuclear weapons, and advocate confidence building measures such as promotion of no first use of nuclear weapons, measures to reduce false alerts and alarms, etc.

In May 1998, India conducted a series of five nuclear weapons tests (Operation "Shakti"), including one thermonuclear device. Earlier moves to conduct such tests in 1995 had been aborted by pressures from the US Clinton administration. This time the tests came as a surprise even to the US intelligence agencies. These tests were followed by a series of tests conducted by Pakistan in the Chagai Hills. International reaction was severe, and prospects of a nuclear war between India and Pakistan were widely discussed. The nuclear embargo on India was tightened, including economic pressures aimed at curbing India's purely indigenous nuclear programme. In contrast Pakistan which had developed its nuclear weapons through the clandestine A.Q. Khan's nuclear smuggling enterprise, plus weapon designs data and political support from China, was relatively unfettered.

India's response to international pressure has been carefully calibrated. It continued to support the total abolition of nuclear weapons, a goal which the US had initially described as "unrealistic", but which now finds some support under President Obama. India has declared a unilateral moratorium on nuclear testing but continues to stay out of the CTBT which it regards as a part of a discriminatory NPT regime. India supports a no first use policy on nuclear weapons, despite the existence of troubled relations with two of its neighbours - China and Pakistan both of which do not declare a no first use policy on nuclear weapons. If it were possible, India could be ready to sign the NPT but as a nuclear weapons state. Nevertheless, India has declared it will respect the "principles" contained in the NPT, while not signing it.

Persistent efforts by India yielded positive results with the July 2005 India –US joint statement on separation of India's civil –US and strategic nuclear programmes, the former to be placed under international safeguards, and in exchange India would benefit from full civil nuclear cooperation. Over the next three years intensive negotiations and discussions with internal constituents in both countries led to amendment of U.S. domestic law, a civil-military nuclear Separation Plan in India, an India-IAEA safeguards (inspections) agreement and the grant of an exemption for India by the Nuclear Suppliers Group. The IAEA India specific safeguards agreement placed some 35 Indian nuclear installations under safeguards in a phased manner.

The joint effort by India and the US to get a waiver from the Nuclear Suppliers Group, in the face of opposition from several countries, must be seen as a landmark in Indo-US diplomatic cooperation. The 45-nation NSG granted the waiver to India on September 6, 2008 allowing it to access civilian nuclear technology and fuel from other countries. The implementation of this waiver makes India the only known country with nuclear weapons which is not a party to the Non Proliferation Treaty (NPT) but is still allowed to carry out nuclear commerce with the rest of the world. India's responsible stewardship of nuclear technology and its declaration on nuclear testing helped this process.

Pakistan, with Chinese support, has lobbied strenuously for the same status from the NSG, but with little success, perhaps due to its involvement in nuclear smuggling. A China-Pakistan nuclear deal, on the lines of the Indo-US is unlikely to gain acceptance from the NSG. Israel, the other undeclared nuclear weapons state with a formidable arsenal, has its own strategic imperatives. The case of North Korea and Iran is quite different, as both these have signed the NPT. North Korea has withdrawn from the NPT and tested a nuclear device. Iran continues to pursue Uranium enrichment and a heavy water reactor, ostensibly for peaceful purposes, but these also give it a nuclear weapons option. Meanwhile the NPT review conference of 2010 did not break any new ground and failed to meet Arab concerns over the problem of Israeli nuclear weapons.

India continues to face challenges in the field of nuclear policy. It is under pressure to sign and ratify the Comprehensive Test Ban Treaty (CTBT). Moves to negotiate a Fissile Material Cutoff Treaty (FMCT) gained momentum after the Obama administration recently changed the US stance on verification, but Pakistan has blocked progress. The FMCT would seek to prohibit the further production of fissile material for nuclear weapons or other explosive devices. This treaty would be difficult to accept unless the goal of a credible nuclear deterrent is achieved. In the future the FMCT negotiations could pose a challenge for India, especially as pressure builds up to move ahead with the Treaty. In the area of civil nuclear cooperation, India is now able to import Uranium fuel for its civilian reactors, which have been run at low output due to fuel shortages. In some countries such as Australia and Japan, there are pressures to insist that India should join the NPT as a condition for civil nuclear commerce.

The Civil Liability for Nuclear Damage Act, 2010 approved by Parliament is seen as a step forward in facilitating civil nuclear commerce, especially with the USA, although some concerns have been raised over the legislation. Recently there has been progress in meeting the specific concerns of nuclear suppliers and vendors over liability. But nuclear power sector in India remains largely restricted to the government sector, and the question is whether this model will be able to manage the financial and technical resources for implementing India's ambitious nuclear power programme.

India's nuclear power programme is based on natural Uranium fuelled pressurized heavy water reactors (PHWR) of the Canadian CANDU design. This type of reactor has been developed by the Indian Atomic Energy establishment and 18 reactors of this type are being operated by the state owned Nuclear Power Corporation of India (NPCIL). In addition 2 reactors of US design from GE (Boiling water Reactor – BWR) using enriched Uranium , and one VVER-1000 reactor of Russian design ( Pressurized water reactor-PWR) using enriched Uranium fuel are operating. 13 reactors have been placed under international safeguards and use imported fuel.. The total nuclear power capacity is 5.3 GigaWatt (GWe) as on April 2016[27]. Four more PHWRs are under construction as well as one more VVER-1000 Russian PWR and one prototype fast breeder reactor (PFBR) at Kalpakkam. The PFBR is expected to lead to a Thorium fuel based generation of commercial fast breeder reactors, a hitherto untried technology. It is planned to have 14.6 GWe nuclear capacity on line by 2024 and 63 GWe by 2032 and to supply 25% of electricity from nuclear power by 2050. India is investing considerable efforts in fast breeder reactors and a thorium based fuel cycle. India will reprocess the used fuel from its safeguarded reactors to recover plutonium for its indigenous three-stage program, using a purpose-built and safeguarded Integrated Nuclear Recycle Plant.

India's nuclear liability law, enacted in 2010[28], reflects India's bitter experience of the Bhopal Gas Tragedy of 1984. Public opinion has been strongly in favour of best possible protection in the case of defective equipment or services by suppliers. Section 17(b) of the new law grants the operator the right to seek recourse from suppliers (only after the operator compensates victims) if the accident was the result of a patent or latent defect in equipment or substandard services. This provision is fundamentally different from the international nuclear industry[29], adds additional risks and uncertainties, and has inhibited private participation in India's nuclear power programme[30].

To get around such concerns, India has ratified the Convention on Supplementary Compensation for Nuclear Damages (CSC) of the IAEA in 2016. In June 2015, the General Insurance Corporation of India (GIC-Re), with other Indian insurance companies, launched the Indian Nuclear Insurance Pool (INIP) with a corpus of Rs. 15 billion. The losses or profit in the pool will be shared by the insurance companies depending on their pre-determined risk capacities. Efforts to get one foreign insurance player, the British Nuclear Insurance Pool to participate did not succeed.

National Power Corporation of India Ltd (NPCIL), the government-owned nuclear power generation company, received India's first insurance policy issued compliant with the guidelines specified in the much discussed Civil Liability for Nuclear Damage Act, 2010. The insurance policy covers all of NPCIL's atomic power plants with a total premium of around Rs. 1 billion for a risk cover of Rs. 15 billion, according to media reports. It remains to be seen whether these steps will succeed in removing obstacles to civil nuclear cooperation with suppliers especially in the US. Indications are that Russia is pursuing additional nuclear power plants in collaboration with NPCIL. A US official stated[31] that the situation had improved and that a commercial agreement with Westinghouse would materialize in 2017.

## Nuclear Security

US President Obama launched an initiative to primarily address the issue of nuclear terrorism. The first Nuclear Security Summit was held in 2010 in Washington , and further Summits were held in 2012 in Seoul,  in 2014 in The Hague, and most recently in Washington in April 2016. The Summits have focused on the need to secure nuclear material and thus prevent nuclear terrorism. Over 50 countries have been participating in these Summits. Under the NSS process, countries work to improve their nuclear security on the basis of the Washington Work Plan, which contains numerous measures and action points. In Seoul a number of additional action points were formulated and set down in the Seoul Communiqué. However, the non participation of Russia, North Korea, Belarus and Iran in the nuclear security summits is significant.

Various countries, including Kazakhstan and Poland, undertook to reduce their highly-enriched uranium stockpiles. Japan agreed to ship additional separated plutonium to the U.S. Canada pledged $42 million to bolster nuclear security. The U.S. disclosed its own inventory of highly enriched uranium has dropped from 741 metric tons in the 1990s to 586 metric tons as of 2013. A strengthened nuclear security agreement, which had languished since 2005, was finally approved, extending safeguards for nuclear materials and requiring criminal penalties for nuclear smuggling. According to the U.S., since the last summit in 2014, ten nations have removed or disposed of about 450 kilograms of highly enriched uranium; Argentina, Switzerland and Uzbekistan are now free of highly enriched uranium, as is all of Latin America and the Caribbean. The summit participants stated that the 2016 summit would be "the last of this kind", indicating that future work would be under the relevant international agencies, notably the IAEA, which would undertake follow on actions in the area of nuclear security.

A notable landmark was the fulfillment of conditions for the entry into force of the 2005 Amendment to the Convention on Physical Protection of Nuclear Materials[32]. This amendment sets forth obligations for states parties to secure their civilian nuclear material -- in use, storage, or transport -- in a manner consistent with International Atomic Energy Agency (IAEA) guidance, and facilitates the further criminalization and prosecution of nuclear smuggling. Among other steps, it also establishes responsibilities for states parties to notify others of potentially dangerous incidents regarding nuclear material out of regulatory control. One troubling scenario is the prospect of non state actors being able to get hold of unsecured radioactive materials and use them as a radiological weapon to inflict civilian casualties. With the increase in inventory of such materials in most countries, the issue of security has gained prominence.

India has taken a high profile in the NSS process and Prime Minister Narendra Modi participated in the 2016 Summit. India announced it would be joining three "gift baskets" or joint endeavours in priority areas[33] – countering nuclear smuggling, the contact group in Vienna to carry on the work of the summit, and sharing best practices through centres of excellence. India has an important stake in nuclear security, given the rise of non-state actors in Pakistan which could gain access to nuclear weapons and materials, as well as security of India's own rapidly growing nuclear programme. India's Prime Minister Narendra Modi announced that India plans to host an international conference with Interpol, a key player in preventing the smuggling of nuclear, biological, radiological and chemical materials. India has offered to host a meeting of Global Initiative to Combat Nuclear Terrorism in 2017.

India's national progress report underlines the various steps the country has taken on nuclear security – updating export controls for companies manufacturing nuclear technology, taking "robust strides" towards implementing nuclear safeguards, setting up an inter-ministerial counter-smuggling team, using low-enriched uranium instead of high-enriched uranium (HEU) and shutting down the only reactor using HEU, setting up 23 response centres across the country to take care of any nuclear or radiological emergency and putting a cyber security architecture in place. India has also announced a $1 million grant for the International Atomic Energy Agency, the lead organization invested in strengthening nuclear security, in addition to the $1 million it contributed in 2013.

## Nuclear Energy

Nuclear power has been in the forefront of practical applications of nuclear technology. The promise of cheap, inexhaustible energy from nuclear fission and fusion has attracted enormous research and development effort over decades. Nuclear power plants using fission have been developed to generate power for electric grids, as well as to propel naval warships and submarines. A wide variety of nuclear reactors have been developed, using different configurations of fissile fuel rods, moderators, coolants and heat exchange systems. The moderator slows down the fast neutrons produced during fission making them more effective in fission. The coolant serves to remove the heat generated in the core of the reactor and convert it into steam in heat exchangers for turning the generators. Reactors using enriched Uranium as fuel usually use water as moderator and coolant, but graphite moderated and gas cooled systems also exist. Reactors using natural Uranium as fuel usually use heavy water as moderator and coolant.

Beginning with the first commercial nuclear power stations which started operation in the 1950s, today there are about 440 commercial nuclear power reactors operating in 31 countries[34], with over 380 GWe (Giga watt electric output[35]) of total capacity. Civil nuclear power can now boast over 16,500 reactor years of experience. About 65 more reactors are under construction while over 150 are firmly planned to be built, equivalent to nearly half of present capacity. They provide over 11% of the world's electricity as continuous, reliable base-load power, without carbon dioxide emissions. In addition, 56 countries operate a total of about 240 research reactors and a further 180 nuclear reactors power some 140 ships and submarines. China, India, Russia, are areas where nuclear power is expanding considerably. In Europe and US the industry is also picking up.

Nuclear power plants have their risks. Accidents may be severely damaging to the environment, especially if they release radioactive substances. A particularly severe accident took place in Chernobyl, Ukraine on 26 April 1986, when a graphite cooled reactor went out of control and released large amounts of radioactivity into the environment[36] which spread far beyond national borders. Other major nuclear accidents were at Three Mile Island, USA[37], and

Fukushima, Japan[38]. Apart from accidents, nuclear reactors during normal operations may release small amounts of radioactive gases and materials which require careful monitoring. Employees working in such plants may be exposed to harmful radiation, which needs monitoring. The reactors produce large amounts of spent fuel which is highly radioactive and needs to be stored safely for long periods, after which the radioactive waste has to be segregated and stored in repositories for hundreds of years. The safety of the reactor depends on several critical systems which could develop faults, leading to major accidents. Operator errors can also lead to accidents, as happened at Chernobyl. Nuclear reactors have a useful lifetime of some 40-50 years, after which they have to be safely decommissioned and the site isolated from the environment.

The economics of nuclear power are also different from conventional fossil fuel plants. Nuclear power plants require high capital investment, and long periods for construction, typically some 5-7 years. The final cost of power depends considerably on financing costs (interest rates), and relatively less on fuel costs[39]. If external costs including cost of Carbon emissions are included, nuclear power could become more competitive with other sources of power.

Even though the nuclear industry has a very good safety record, the high profile nature of the accidents at Chernobyl and Fukushima have resulted in growing public concern and opposition to nuclear power. In India, there has been considerable public opposition to nuclear plants being set up at Kudankulam in Tamil Nadu, as well as at Jaitapur, Maharashtra. Both locations are near the sea and opposition has emerged from local residents, fishermen and from concerns over possible tsunami impact. In this debate perceptions may sometimes outweigh scientific facts and reasoning.

Nuclear reactor technology is an area of active research and development. New types of fission reactors with improved safety and efficiency are being researched, including fast neutron reactors. An international collaboration, the GIF was set up in 2001[40] to explore new generations of reactors. There are currently ten active members of the Generation IV International Forum (GIF): Canada, China, the European Atomic Energy Community (Euratom), France, Japan, Russia, South Africa, South Korea, Switzerland, and the United States. The GIF has identified six reactor technologies for future development. These were selected on the basis of being clean, safe and cost-effective means of meeting increased energy demands on a sustainable basis, while being resistant to diversion of materials for weapons proliferation and secure from terrorist attacks. Expenditure of about $6 billion over 15 years is planned on development. About 80% of the cost is being met by the USA, Japan and France.

India is not involved with the GIF but is developing its own advanced technology to utilize thorium as a nuclear fuel. A three-stage program has the first stage well-established, with pressurized heavy water reactors (PHWR) fuelled by natural uranium to generate plutonium. Then fast breeder reactors (FBRs) use this plutonium-based fuel to breed U-233 from thorium, and finally advanced nuclear power systems will use the U-233. The spent fuel will be reprocessed to recover fissile materials for recycling. The two major options for the third stage, while continuing with the PHWR and FBR programmes, are an Advanced Heavy Water Reactor and subcritical Accelerator-Driven Systems.

An accelerator-driven subcritical (ADS) reactor is a nuclear reactor design formed by coupling a substantially subcritical nuclear reactor core with a high-energy proton accelerator. It would use Thorium as a fuel, which is more abundant than uranium. The neutrons needed for the fission process would be provided by the particle accelerator and enabling fission without needing to make the reactor critical. Such reactors have several advantages such as inherent safety, and the relatively short life of its waste products. The high energy proton beam impacts a molten lead target inside the core, chipping or "spallating" neutrons from the lead nuclei. These spallation neutrons convert fertile thorium to fissile uranium-233 and drive the fission reaction in the uranium. Thorium produces 200 times more power per kilogram than uranium. Further, thorium reactors can generate power from the plutonium residue left by uranium reactors. Thorium does not require significant refining, unlike uranium and has a higher neutron yield per neutron absorbed. This type of reactor is inherently safe, since thorium is not fissile, and the fission process stops when the proton beam stops, as when power is lost, thus the reactor is subcritical. Only microscopic quantities of plutonium are produced and can be burned in the same reactor. A Norwegian company, Aker Solutions is developing a small 600 MWe thorium reactor to be located underground suitable for supplying small grids and which does not require an enormous facility for safety and security. Costs for the first reactor are estimated at about $2.4 billion.

India has a big stake in reactor technology using Thorium since it has one of the world's largest reserves of this material. In view of this India is actively engaged in cutting edge research into reactor technology that can use Thorium. A facility for carrying out research into ADS has been set up at BARC. A 300 MWe Advanced Heavy Water Reactor (AHWR)[41] prototype is being developed at BARC, India, for the purpose of thorium utilization, as part of India's nuclear power programme. It uses heavy water as moderator, light water as coolant, and Thorium fuel in a configuration similar to the PHWRs used in India. If successful, it could bring down the capital cost of nuclear power and improve safety and security.

## Fusion Energy

Energy from nuclear fusion has been a dream for mankind. Controlled fusion could produce power in abundance, using the very fusion reactions that power the stars like the sun. An ambitious multinational project to harness and control nuclear fusion is ongoing, the International Thermonuclear Experimental Reactor (ITER) located in Cadarache, France[42]. The ITER fusion reactor has been designed to produce 500 megawatts of output power for several seconds while needing 50 megawatts to operate. Thereby the machine aims to demonstrate the principle of producing more energy from the fusion process than is used to initiate it. The project is funded and run by seven member entities—the European Union, India, Japan, China, Russia, South Korea and the United States. The EU, as host party for the ITER complex, is contributing about 45 percent of the cost, with the other six parties contributing approximately 9 percent each. Construction of the ITER Tokamak complex started in 2013 and the building costs are now over US$14 billion as of June 2015. The facility is expected to finish its construction phase in 2019 and will start commissioning the reactor that same year and initiate plasma experiments in 2020 with full deuterium–tritium fusion experiments starting in 2027.

A parallel effort at harnessing fusion energy is the Stellarator, a device that uses magnetic fields in a complicated twisted geometry to confine the hot plasma for sufficiently long to achieve fusion. While the Tokamak is basically operated in a pulsed mode, the Stellarator is capable of operating in a continuous mode but is far more complicated to construct. The most advanced Stellarator project is the Wendelstein-7X[43], at Griefswald, Germany, operated by the Max Planck Institute of Plasma Physics. Over 18 technical institutions from Germany, Europe, Japan and US are collaborating in this project. The main assembly of Wendelstein 7-X was concluded in 2014. Once all technical systems had been checked step by step the first plasma was produced on 10th December 2015.

The growing trend to international collaboration in nuclear reactor research is quite striking. Indeed collaboration is vital to ensure the best possible safety, security and efficiency of nuclear power plants. In addition collaboration can spread best practices in training, operations, and maintenance, and regulatory mechanisms for the mutual benefit of mankind. Such collaboration can also serve to build confidence and bridge strategic and economic rivalries.

## Endnotes

1  "Little Boy" was a gun type fission weapon using 64 kg of enriched Uranium. It was detonated on 6 Aug 1945 over the city of Hiroshima. weighed 4.4 tonnes and released energy of 15 kilotonnes of TNT. 32 such devices were manufactured at the Los Alamos Laboratory. These relatively inefficient devices were withdrawn in 1951.

2  "Fat Man" was an implosion type nuclear weapon which used 6.2 kg of Plutonium. It weighed 4.6 tonnes. It was detonated over the city of Nagasaki on 9 Aug 1945, releasing an energy of 21 kilotonnes of TNT. It was first tested at Alamogordo on 16 July 1945 named the "Trinity test". 120 such devices were made at the Los Alamos Laboratory during 1945-49. The weapon was retired in 1950.

3  The United States conducted six atomic tests before the Soviet Union developed their first atomic bomb (RDS-1) and tested it on August 29, 1949. By the 1950s the United States had established a dedicated test site on its own territory (Nevada Test Site) and was also using a site in the Marshall Islands (Pacific Proving Grounds) for extensive atomic and nuclear testing. The Soviet Union also began testing on a limited scale, primarily in Kazakhstan. During the later phases of the Cold War, though, both countries developed accelerated testing programs, testing many hundreds of bombs over the last half of the 20th century.

4  Statement to Indian Parliament (Lok Sabha), 2 April 1954. He said "nuclear, chemical and biological energy and power should not be used to forge weapons of mass destruction". He called for negotiations for prohibition and elimination of nuclear weapons and in the interim, a standstill agreement to halt nuclear testing. This call was not heeded.

5  Tsar Bomba is the nickname for the AN602 hydrogen bomb, the most powerful nuclear weapon tested on October 30, 1961. It had a yield of 50 megaton TNT. Only one bomb of this type was ever officially built and it was detonated in the Novaya Zemlya archipelago at Sukhoy Nos. The bomb, weighed 27 metric tons, 8 metres long by 2 metres in diameter. The Tsar Bomba was a three-stage bomb with a yield of 50 to 58 megatons of TNT. The initial three-stage design was capable of yielding approximately 100 Mt, but it would have caused too much nuclear fallout and the plane delivering the bomb would not have

enough time to escape the explosion. To limit fallout, the third stage and possibly the second stage yields were reduced.

6   The Partial Test Ban Treaty was signed and ratified by the governments of the Soviet Union, the United Kingdom, and the United States during 1963. By 2013, 125 UN member states had ratified or acceded to the treaty, and a further 10 states had signed but not ratified the treaty. It prohibits all test detonations of nuclear weapons except underground. Countries known to have tested nuclear weapons but which have not signed the treaty are China, France and North Korea.

7   Article XIV of the Treaty states that it will enter into force after the following 44 States listed in Annex 2 to the Treaty have ratified it: Algeria, Argentina, Australia, Austria, Bangladesh, Belgium, Brazil, Bulgaria, Canada, Chile, China, Colombia, Democratic People's Republic of Korea, Egypt, Finland, France, Germany, Hungary, India, Indonesia, Iran (Islamic Republic of), Israel, Italy, Japan, Mexico, Netherlands, Norway, Pakistan, Peru, Poland, Romania, Republic of Korea, Russian Federation, Slovakia, South Africa, Spain, Sweden, Switzerland, Turkey, Ukraine, United Kingdom of Great Britain and Northern Ireland, United States of America, Viet Nam, Zaire.

8   On 10 September 1996, the General Assembly adopted a draft resolution, initiated by Australia and sponsored by 126 States, by a vote of 158 in favour, 3 against (Bhutan, India, Libya), with 5 abstentions (Cuba, Lebanon, Mauritius, Syria, Tanzania), and thereby adopted the Comprehensive Nuclear-Test Ban Treaty and requested the Secretary-General of the United Nations, in his capacity as Depositary of the Treaty, to open it for signature at the earliest possible date.

9   As of September 2015, 104 countries are parties to the Outer Space Treaty, while another 24 have signed the treaty but have not completed ratification. The follow up Moon Treaty of 1979 failed to be ratified by any major space-faring nation such as those capable of orbital spaceflight.

10  As of May 2013, 94 current states are parties to the treaty, while another 21 have signed the treaty but have not completed ratification.

11 A total of 191 states have joined the Treaty, though North Korea, which acceded to the NPT in 1985 but announced its withdrawal in 2003. Four UN member states have never joined the NPT: India, Israel, Pakistan and South Sudan.

12 A key factor for India's acquisition of nuclear weapons is the potential threat presented by the nuclear-armed state of China, which faces India along much of its northern border. Disputes covering 80,000 square kilometers of this border region exist, and a war with China in October 1962 resulted in a defeat for Indian forces.

13 Prime Minister Z A Bhutto remarked in 1965 that "If India builds the bomb, we will eat grass and leaves for a thousand years, even go hungry, but we will get one of our own. The Christians have the bomb, the Jews have the bomb and now the Hindus have the bomb. Why not the Muslims too have the bomb?" In January 1972, President Zulfikar Ali Bhutto committed Pakistan to acquiring nuclear weapons at a secret meeting held in Multan in the wake of the country's devastating defeat in the 1971 Bangladesh war.

14 Between 0 and 60 degrees South latitude and East and West of the 115 degrees East meridian

15 Protocol I contains assurances of non use of nuclear devices against parties, while Protocol II contains obligations not to test nuclear devices in the territory of the parties.

16 Arab countries, led by Egypt, and Iran have since the mid-1970s sought, unsuccessfully, to pressure Israel to dismantle its nuclear arsenal and establish a Middle East nuclear-weapon-free zone (MENWFZ). But they have little leverage to pressure Israel to change its nuclear policy and US support for Israel's nuclear monopoly in the region remains as strong as ever. "Prospects for a Nuclear-Weapon-Free Zone in the Middle East", Gawdat Bahgat, ISN, 15 June 2015 http://www.isn.ethz. ch/Digital-Library/Articles/Detail/?lng=en&id=191392 accessed 26-3-2016

17 The number of nuclear weapons in the region is estimated at China (250), India(120) and Pakistan (130), in addition to stocks of fissile material. All three are developing land, sea and air delivery systems.

Southern Asia's Nuclear Powers, Eleanor Albert, CFR 9 November 2015 http://www.cfr.org/asia-and-pacific/southern-asias-nuclear-powers/ p36215 , accessed 26-3-2016

18 A Nuclear Weapon-Free Zone in Europe, Peace Research Institute, Frankfurt, 2010, https://www.bmeia.gv.at/fileadmin/user_upload/ Zentrale/Aussenpolitik/Abruestung/NWFZE_Finalversion.pdf accessed 26-3-2016

19 Fissile Material Cut Off Treaty, Reaching Critical Will, http://www. reachingcriticalwill.org/resources/fact-sheets/critical-issues/4737-fissile-material-cut-off-treaty , accessed 23-4-2016

20 Fissile Material Cut-off Treaty (FMCT) at a Glance, Aug 2013, Arms Control Asssociation, https://www.armscontrol.org/factsheets/fmct , accessed 23-4-2016

21 On 27 October 2016, the First Committee of the UN General Assembly adopted resolution L.41 to convene negotiations in 2017 on a "legally binding instrument to prohibit nuclear weapons, leading towards their total elimination". The voting result was 123 nations in favour and 38 against, with 16 abstentions. India, Pakistan and China abstained.

22 UN General Assembly approves historic resolution, ICAN, December 23, 2016, http://www.icanw.org/campaign-news/un-general-assembly-approves-historic-resolution . accessed 27-1-2017

23 The New Agenda Coalition (NAC), composed of Brazil, Egypt, Ireland, Mexico, New Zealand and South Africa, is a geographically dispersed group of middle power countries seeking to build an international consensus to make progress on nuclear disarmament, as legally called for in the nuclear NPT.

24 A Treaty to Ban Nuclear Weapons Is in the Making,Sergio Duarte, Inter Press Service, 14 Apr 2017, http://www.ipsnews.net/2017/04/a-treaty-to-ban-nuclear-weapons-is-in-the-making/ , accessed 19-4-2017

25 The Day A Nuclear Conflict Was Averted, Strobe Talbott, 13 Sept 2004, Yale Global, http://yaleglobal.yale.edu/content/day-nuclear-conflict-was-averted , accessed 20-4-2017

26  No First Use Over Non-Testing: Practical Options for South Asian Regional Stability, Reshmi Kazi, 9 Sept 2016, IDSA News, http://www.idsa.in/idsanews/no-first-use-over-non-testing ,accessed 20-4-2017

27  Nuclear Power in India, World Nuclear Association, April 2016, http://www.world-nuclear.org/information-library/country-profiles/countries-g-n/india.aspx  accessed 25-4-2016

28  India joins nuclear liability convention, 5 Feb 2016, World Nuclear News, http://www.world-nuclear-news.org/NP-India-joins-nuclear-liability-convention-0502167.html  accessed 24-4-2016

29  The US has adopted the Price Anderson Act, 1957, the world's first comprehensive nuclear liability law. It now provides $13.6 billion in cover without cost to the public or government and without fault needing to be proven. It covers power reactors and all other nuclear facilities and was renewed for 20 years in mid 2005. It requires individual operators to be responsible for two layers of insurance cover. Under the Layer 1 each nuclear operator is required to purchase US$ 450 million liability cover (from 2017) for each reactor which is provided by a private insurance pool, American Nuclear Insurers (ANI). This is financial liability, not legal liability as in European liability conventions. Layer 2 or secondary financial protection (SFP) program is jointly provided by all US reactor operators. It is funded through retrospective payments if required of up to $121 million per reactor per accident collected in annual instalments of $19 million (and adjusted with inflation). Combined, the total provision in 2014 from both layers comes to over $13.6 billion paid for by the utilities. Beyond this cover and irrespective of fault, the US Congress, as insurer of last resort, must decide how compensation is provided in the event of a major accident.

30  Private sector participation in nuclear power is not presently permitted by the Government. However the private sector (domestic and foreign) is involved as suppliers of equipment and services to government nuclear entities. The liability law would therefore impact them.

31  US hopeful of commercial deal for nuclear plant in India by 2017, Economic Times, 16 Jan 2017, http://economictimes.indiatimes.com/news/politics-and-nation/us-hopeful-of-commercial-deal-for-nuclear-

plant-in-india-by-2017/printarticle/56594182.cms , accessed 29-3-2017

32 Statement by National Security Advisor Susan E. Rice on the Entry Into Force of the 2005 Amendment to the Convention on Physical Protection of Nuclear Materials April 14, 2016, http://www.nss2016. org/news/2016/4/14/statement-by-national-security-advisor-susan-e-rice-on-the-entry-into-force-of-the-2005-amendment-to-the-convention-on-physical-protection-of-nuclear-materials , accessed 24-4-2016

33 Indian Diplomacy in Full Flow at Nuclear Security Summit, Eyes Firmly Set on NSG Next, Seema Sirohi, The Wire, 2 April 2016, http://thewire.in/2016/04/02/indian-diplomacy-in-full-flow-at-nuclear-security-summit-27135/ accessed 24-4-2016

34 Nuclear Power in the World Today, World Nuclear Association, January 2016, http://www.world-nuclear.org/information-library/current-and-future-generation/nuclear-power-in-the-world-today.aspx , accessed 24-4-2016

35 Giga Watt electric (GWe) is one billion watts of electrical power. A nuclear power plant typically converts only about 40 percent of the heat energy produced ion the reactor into useful electrical output, the rest has to be released into the environment. Reactor output is given in terms of thermal output ( MWth) or electrical output (MWe)

36 The Chernobyl accident of 24 April 1986 was the result of a flawed reactor design that was operated with inadequately trained personnel. During a test the Chernobyl-4 reactor, a Soviet designed graphite moderated water cooled reactor (RBMK-1000) became unstable and could not be shut down. The resulting steam explosion and fires released at least 5% of the radioactive reactor core into the atmosphere and downwind – some 5200 PBq (I-131 eq). Two Chernobyl plant workers died on the night of the accident, and a further 28 people died within a few weeks as a result of acute radiation poisoning. The accident released large quantities of radioactive substances into the air for about 10 days, affecting large populations in Belarus, Russia and Ukraine. and to some extent over Scandinavia and Europe. Chernobyl Accident, 1986, World Nuclear Association, http://www.world-nuclear.org/information-

library/safety-and-security/safety-of-plants/chernobyl-accident.aspx accessed 25-4-2016.

37 The Three Mile Island Unit 2 (TMI-2) reactor, near Middletown, PA, USA, partially melted down on March 28, 1979. This was the most serious accident in U.S. commercial nuclear power plant operating history, although its small radioactive releases had no detectable health effects on plant workers or the public. Its aftermath brought about sweeping changes involving emergency response planning, reactor operator training, human factors engineering, radiation protection, and many other areas of nuclear power plant operations. It also caused the NRC to tighten and heighten its regulatory oversight. All of these changes significantly enhanced U.S. reactor safety. A combination of equipment malfunctions, design-related problems and worker errors led to TMI-2's partial meltdown and very small off-site releases of radioactivity. Backgrounder on the Three Mile Island Accident, Nuclear Regulatory Commission USA, http://www.nrc.gov/reading-rm/doc-collections/fact-sheets/3mile-isle.html , accessed 25-4-2016

38 Following a major earthquake, a 15-metre tsunami disabled the power supply and cooling systems of three Fukushima Daiichi reactors, Japan, causing a nuclear accident on 11 March 2011. All three reactor cores largely melted in the first three days. There were high radioactive releases over days 4 to 6, eventually a total of some 940 PBq (I-131 eq).After two weeks, the three reactors (units 1-3) were stabilized with water addition and official 'cold shutdown condition' was announced in mid-December. Apart from cooling, the basic ongoing task was to prevent hydrogen explosions, and release of radioactive materials, particularly in contaminated water leaked from the three units. There have been no deaths or cases of radiation sickness from the nuclear accident, but over 100,000 people were evacuated from their homes to ensure this. Government nervousness delays the return of many. Fukushima accident, World Nuclear Association, April 2016, http://www.world-nuclear.org/information-library/safety-and-security/safety-of-plants/fukushima-accident.aspx , accessed 25-4-2016

39 At a 10 percent discount rate, the levelised cost of nuclear electricity generation in OECD countries range between 4.2 US cents/KWh (Korea) and 13.7 US Cents/KWh (Switzerland). The share of

investment in total levelised generation cost is around 75% while the other cost elements, operation and maintenance costs and fuel cycle costs, represent 15% and 9% respectively. These figures also include costs for refurbishment, waste treatment and decommissioning after a 60-year lifetime. Press kit: Economics of nuclear power, Nuclear Energy Agency, https://www.oecd-nea.org/news/press-kits/economics.htm , accessed 25-4-2016

40 Generation IV Nuclear Reactors, World Nuclear Association, http://www.world-nuclear.org/information-library/nuclear-fuel-cycle/nuclear-power-reactors/generation-iv-nuclear-reactors.aspx , accessed 26-4-2016

41 Advanced Heavy Water Reactor, BARC, India, http://www.barc.gov.in/reactor/ahwr.pdf , accessed 30-1-2017

42 What is ITER? ITER website, https://www.iter.org/proj/inafewlines , accessed 26-4-2016

43 Wendelstein 7-X, Max Planck Institute for Plasma Physics, http://www.ipp.mpg.de/16900/w7x , accessed 26-4-2016

# Chapter 5

# Chemical Technology – Progress or a Plague?

*"Chlorine is a deadly poison gas employed on European battlefields in World War I. Sodium is a corrosive metal which burns upon contact with water. Together they make a placid and unpoisonous material, table salt. Why each of these substances has the properties it does is a subject called chemistry."*

*— Carl Sagan, 1979*

The 20th century started with fundamental advances in chemistry, drawing on developments in modern physics regarding atomic structure and quantum mechanics. This led to greater understanding of the forces that bind atoms together, the so called "chemical bond", as well as an explanation for the periodic table of elements, discovered in 1869. The chemical bond was understood in terms of sharing of outer electrons of the participating atoms, according to the principles of quantum mechanics. Industrial chemistry developed rapidly from the 19th century, producing items such as dyes, bleaching powder and sulphuric acid for use in the textile industry, and chemicals such as caustic soda and chlorine from salt. Production of petroleum based chemicals such as plastics and vegetable oil derived soaps also grew.

The chemical industry grew rapidly in Britain, France, Germany and the USA. By the 1920s the global chemical industry was dominated by large conglomerates in these countries. Germany developed through new technology development, while the US developed by importing and adapting technology from Europe[1]. By 1913 Germany dominated the global chemical industry.

Chemical research progresses broadly along three broad tracks. Firstly the discovery of and finding useful applications of substances found in nature and ways to produce such substances. Secondly, the discovery of new and more efficient ways of producing useful substances. Thirdly, the discovery of substances not found in nature with useful applications and to exploit them. Since the 19th century there has been rapid progress along all these tracks. Much of innovation in this field is concerned with scaling up a process from laboratory or pilot scale to an industrial scale. This has led to the development of the discipline of chemical engineering. Advances in this field have led to the development of thousands of synthetic materials that have made possible the establishment of new industries. Bakelite, the earliest synthetic resin, was produced in 1909. Since then, plastics such as polyethylene, Teflon, Lucite, and the silicones have found wide-spread application. Drugs such as the sulfa drugs, penicillin, antihistamines, and antibiotics have also been synthesized.

With the start of the First World War in August 1914, the German chemicals industry faced serious challenges. Companies could no longer import the raw materials they urgently needed, and they increasingly shifted their activities to weapons production[2]. Shortages of imported raw materials had to be overcome by developing alternative chemical processes for explosives, munitions and fertilizers. In both World Wars, shortages of natural raw materials led to development of products such as synthetic rubber, and gasoline from coal.

## Chemical Warfare

The German war ministry decided during World War I to use chlorine gas as a weapon, in violation of the Hague Conventions which banned the use of poisonous gas. The German army first used chlorine gas on April 22, 1915, at the Belgian town of Ypres.

The surprise use of chlorine gas allowed the Germans to rupture the French line along a 6-kilometer front, causing terror and forcing a panicked and chaotic retreat[3]. Within a matter of minutes, this slow moving wall of gas killed more than 1000 French and Algerian soldiers, while wounding approximately 4000. The German High Command sanctioned the use of gas in the hope that this new weapon would bring a decisive victory by breaking the enduring stalemate of trench warfare. However, having no plan to send a large offensive force in after the gas, the Germans were unable to take advantage of the situation. The attack marked a turning point in military history, as it is recognized as the first successful use of lethal chemical weapons on the battlefield.

Estimated casualties from the use of poison gas during the World War I as a whole were more than one half-million soldiers wounded and 15,000 to 20,000 killed among the troops of Germany, Britain, France, and the United States. Russian troops alone are believed to have suffered more than a half-million casualties; due to the poor protective equipment available on the Russian side, it is likely that about 10 percent of these casualties were fatalities, representing an additional 50,000 to 60,000 killed. No statistics are available with regard to civilian casualties from the use of poison gas, or the later consequences of gas-related injuries to soldiers who survived the war.

The German company Bayer developed a series of new poison gas weapons. First came phosgene, which was more lethal than chlorine gas and this was followed by mustard gas. BASF joined Bayer as a major supplier of phosgene and of intermediary products required in the production of mustard gas. The so-called "blue cross" grenades contained a gas that caused vomiting and pulmonary irritation, forcing soldiers to remove their gas masks and breathe in the deadly mustard gas or chlorine gas which would be fired off simultaneously. The chemical industry labored constantly to "optimize" the use of poison gases. Prior to the development of poison gas grenades, chlorine gas was simply released from a tank and blown in the direction of the enemy using fans.

In July 1917, aware of the loss of their technological superiority and perhaps their ability to win the war, the Germans deployed a new and more troublesome chemical agent: mustard gas. Although mustard gas was introduced late in the war, it became known as the "King of Battle Gases" because it eventually caused more chemical casualties than all the other agents combined. Mustard gas was a particular problem for both sides because after it was released it settled in an area, contaminating it, and being heavier than air or water, it settled in ditches or at the bottom of trenches and puddles and created a persistent environmental hazard for troops, civilians, and animals alike. More than 80% of the approximately 186 000 British chemical casualties were caused by mustard gas alone, with a death toll of approximately 2.6%.

By the time of the armistice on November 1918, the use of chemical weapons such as chlorine, phosgene, and mustard gas had resulted in more than 1.3 million casualties and approximately 90 000 deaths. While the Hague Declaration of 1899 and the Hague Convention of 1907 forbade the use of "poison or poisonous weapons" in warfare, yet more than 124 000 tons of gas were produced by the end of World War I. By war's end the national programs among the warring nations focused on both the offensive and defensive aspect of chemical weapons and involved an array of specialists and institutions working in fields such as chemistry, physics, and engineering and, increasingly, from medicine, biology, and physiology, further blurring ethical demarcations in research.

Chemical warfare has remained surprisingly resilient. Between the 1918 armistice and 1933, several international conferences were held to try to limit or abolish chemical weapons. Although progress was made toward outlawing the use of "asphyxiating, poisonous or other gases" per the Geneva Protocol of 1925, programs and research continued throughout the interwar period and most of the rest of the century, despite the public's rejection of these weapons. Most countries refused to eliminate chemical weapons as a strategic weapon, despite condemning them, although all major combatants dramatically scaled back their programs. During World War II, for instance, the US military was vocal about its avoidance of the deployment or use of poison gas. At the same time, however, chemical weapons were a mainstay of the Army Air Corps strategic bombing campaigns. During the Cold War, work on a variety of chemical weapons continued on both sides in offensive weapons as well as defensive measures. Nerve agents such as VX, and a new class of binary weapons were developed. The technology was shared by countries on both sides of the Cold War.

## The Chemical Weapons Convention[4]

In the wake of World War I, during which the world witnessed the horrors of large-scale chemical warfare, international efforts to ban the use of chemical weapons and prevent such suffering from being inflicted again, on soldiers and civilians, intensified. The result of this renewed global commitment was the 1925 Geneva Protocol for the Prohibition of the Use of Asphyxiating, Poisonous or Other Gases, and Bacteriological Methods of Warfare.

The Geneva Protocol did not prohibit the development, production or possession of chemical weapons but only banned the use of chemical and bacteriological (biological) weapons in war. Furthermore, many countries had signed the Protocol with reservations permitting them to use chemical weapons against countries that had not joined the Protocol or to respond in kind if attacked with chemical weapons.

Many developed countries spent considerable resources on the development of Chemical Weapons, particularly after the discovery of powerful nerve gases. Chemical Weapons were used by a number of countries in the inter-war period, and all the major powers involved in World War II anticipated that large-scale chemical warfare would take place. Contrary to expectations, however, Chemical Weapons were never used in Europe in World War II. The reasons could have been fear of retaliation in kind, the level of protection of enemy troops, or moral reasons that deterred their use. The fate of some of the stockpiles built up in anticipation of World War II is also uncertain. Many Chemical Weapons were abandoned, buried or simply dumped at sea. In any event, following World War II, and with the advent of the nuclear debate, several countries gradually came to the realization that the value of having Chemical Weapons in their arsenals was limited, while the threat posed by the availability and proliferation of such weapons made a comprehensive ban desirable.

The issue of Chemical Weapons was on the agenda of the Geneva Conference on Disarmament since 1968 though initially linked with Biological weapons, but separated subsequently. Beginning in 1986, the global chemical industry actively participated in these negotiations. The negotiators reached an understanding that a ban on these weapons would be subject to international verification. With the thawing of relations between the United States and the Soviet Union, progress increased. In 1990, the United States and the Soviet Union signed a bilateral agreement on Chemical Weapons, under which the two countries agreed not to produce Chemical Weapons, and to reduce their stocks. Negotiations succeeded in resolving several remaining issues and the Convention was finally adopted by the UN General Assembly in December 1992.

The Convention entered into force in April 1997, by which time a parallel process of setting up the Organization for the Prohibition of Chemical Weapons (OPCW) had been carried out by a Preparatory Committee. OPCW currently has 192 Member States, which share the collective goal of preventing chemistry from being used for warfare. To this end, the Convention contains four key provisions: (1) destroying all existing chemical weapons under international verification by the OPCW; (2) monitoring chemical industry to prevent new weapons from re-emerging;(3) providing assistance and protection to States Parties against chemical threats; and (4) fostering international cooperation to strengthen implementation of the Convention and promote the peaceful use of chemistry.

## Use of Chemical Weapons and toxic substances

While the CWC has been a success story in terms of number of ratifying countries and the phasing out of chemical arsenals, there are some outstanding problems. Chemical weapons have been used by Iraq[5] during its war with Iran. During the recent conflict in Syria ( 2011 onwards), there have been several instances of reported use of chemical weapons[6] by the parties to the conflict. There is concern over the possible use of chemical agents by non-state actors including terrorist groups and individuals, such as Aum Shinrikyo (responsible for the 1995 Tokyo subway attack using Sarin gas), Al Qaeda (Ricin), Islamic State ISIS (Mustard gas). The easy availability of precursor chemicals and knowledge of chemical processes has made it possible for crude chemical weapons to be produced by terrorists groups and even individuals.

## The Australia Group

The Australia Group is an informal group of countries plus the European Commission established in 1985 (after the use of chemical weapons by Iraq in 1984) at the initiative of Australia, to help member countries to identify those exports which need to be controlled so as not to contribute to the spread of chemical and biological weapons. Australia manages the secretariat. The group, initially consisted of 15 members but has expanded to 42, with the incorporation of Mexico on August 12, 2013. Its members are Organization for Economic Co-operation and Development (OECD) members, the European Commission, and all 28 member states of the European Union, Ukraine, and Argentina.

The initial members of the group had different assessments of which chemical precursors should be subject to export control. Later adherents initially had no such controls. Today, members of the group maintain export controls on a uniform list of 54 compounds, including several that are not prohibited for export under the Chemical Weapons Convention, but can be used in the manufacture of chemical weapons. In 2002, the group took two important steps to strengthen export control. The first was the "no-undercut" requirement, which stated that any member of the group considering making an export to another state that had already been denied an export by any other member of the group must first consult with that member state before approving the export. The second was the "catch-all" provision, which requires member states to halt all exports that could be used by importers in chemical or biological weapons programs, regardless of whether or not the export is on the group's control lists. Delegations representing the members meet every year in Paris, France.

## India's role in the OPCW and the Australia Group[7]

India has an elaborate and comprehensive export controls system for chemicals that could be used for chemical warfare. Strict control of chemicals is considered important because chemicals constitute the largest category of Indian exports. President Obama, during his trip to India, endorsed India's candidature for the Australia Group. Later, France supported the Indian candidature. A team of the Australia Group visited India in the last week of April 2011 and interacted with Indian officials and experts working on export controls.

India has fulfilled all the criteria but putting additional items of the Australia Group in its export controls policy is facing resistance from chemical industry. Putting additional items on the Indian Control list called Special Chemicals, Organisms, Materials, Equipment and Technology List, which is more popular by its abbreviation SCOMET, means additional licensing burden on Indian exporters. The item 1 of the SCOMET list contains special chemical items.

The Indian export control system has impressive legal, institutional and enforcement frameworks. India is a signatory to the CWC. India has consequently incorporated all the three schedules of CWC into its control list, and has passed laws such as the Chemical Weapons Convention Act of 2000. India is further amending the

CWC Act to include some enforcement provisions, especially regarding personnel.

The Directorate General of Foreign Trade is the nodal agency for granting license for SCOMET controlled chemicals. However, all the licenses for such export are referred to an inter-service agency. The Directorate has devised several parameters for scrutiny of license applications.

The amendment in the 2000 CWC Act is in progress to introduce some enforcement related provisions. Once this amendment is passed, enforcement capacity of the nodal agencies will be further strengthened. The chemical industry faces certain problems, such as difficulties in distinguishing a commercial consignment from a chemical weapons-related consignment, establishing appropriate commodity thresholds, personal safety inspectors and the diversified but highly specialized nature of chemicals. All major customs houses in India have their own chemical labs so any suspicious items can be tested, and in addition, if required, at a laboratory of the Defence Research and Development Organization.

## Chemical Industry today

Today , the chemicals industry is one of the largest industries in the world, with global sales in 2010 of more than $3.1 trillion, according to the International Council of Chemical Associations (ICCA). The ICCA also says more than 20 million jobs are directly or indirectly connected to the chemical industry.

From 1945 to 1980, the chemical industry enjoyed substantial growth. But the early 1980s witnessed the most severe decline since the Great Depression of the 1930s. Although companies had developed new, more efficient production methods, skyrocketing petroleum prices greatly reduced world demand for basic chemicals. In order to compete more effectively, many chemical firms spent heavily on expansion and research and development. Important problems still confront chemical industries: the development of alternate sources of energy, elimination of environmental pollution, and improvement of the safety of existing chemical products. Nanotechnology, the study of controlling matter at its most fundamental atomic level, is positioned to redefine the chemical manufacturing process, but it also requires massive research and development.

Another development impacted the chemical industry occurred in the 21st century: the green chemistry movement. This took place when the chemical industry, responding to consumer concerns, worked to develop more environmentally friendly products and reduce its impact on the environment. According to the American Chemical Society, green chemistry is the design, development, and implementation of chemical products and processes that reduce or eliminate the use of chemicals that are hazardous to human health and the environment.

## Chemical Safety

Chemical Safety is achieved by undertaking all activities involving chemicals in such a way as to ensure the safety of human health and the environment. It covers all chemicals, natural and manufactured, and the full range of exposure situations from the natural presence of chemicals in the environment to their extraction or synthesis, industrial production, transport use and disposal. Chemical safety has many scientific and technical components. Among these are toxicology, ecotoxicology and the process of chemical risk assessment which requires a detailed knowledge of exposure and of biological effects.

Chemical releases may arise from accidents in manufacturing, transport or storage facilities, technological incidents, natural disasters, conflicts and terrorism. Examples of chemical accidents are the accident at a Union Carbide plant in Bhopal, India (1984) which caused 3000 deaths, and the ICMESA plant at Seveso, Italy (1976) which released toxic chemicals affecting some 500 persons. To deal with such accidents, the UK's National Chemical Emergency Centre was set up in 1973 to understand the requirements of the chemical industry and emergency services.

The use of pesticides and chemicals in agriculture results in toxic substances being present in food products leading to a rise in diseases such as cancer. Similarly the use of pesticides and agrochemicals  may result in the appearance of persistent toxic products in the environment, including in water and soil. Despite international and national efforts to reduce such chemical pollution it remains a formidable problem affecting food products.


## Endnotes

1  Chemical Industries after 1850, Johann Peter Murmann , Oxford Encyclopedia of Economic History, http://www.professor-murmann. net/murmann_oeeh.pdf accessed 16--11-2015

2  The Role of the German Chemical Industry in the First World War; Wolheim Memorial, http://www.wollheim-memorial.de/en/ chemieindustrie_im_ersten_weltkrieg_en accessed 16-11-2015

3  Chemical Warfare and Medical Response During World War I, Gerald Fitzgerald, Am J Public Health. 2008 April; 98(4): 611–625. http:// www.ncbi.nlm.nih.gov/pmc/articles/PMC2376985/ accessed 16-11-2015

4  Genesis and Historical Development of the chemical Weapons Convention, OPCW,

https://www.opcw.org/chemical-weapons-convention/genesis-and-historical-development/

5  The Iraqi Army initiated (1978–1991) offensive chemical weapons (CW) programs involving mustard gas and sarin. President Saddam Hussein waged chemical warfare against Iran during 1983-88. He also used chemicals in 1988 against his civilian Kurdish population

and during a popular uprising in the south in 1991. About 100,000 Iranian soldiers were victims of Iraq's chemical attacks. Many were hit by mustard gas. The official estimate does not include the civilian population contaminated in bordering towns. In March 1988, the Iraqi Kurdish village of Halabja was exposed to multiple chemical agents dropped from warplanes; these may have included mustard gas, the nerve agents sarin, tabun and VX and possibly cyanide. Between 3,200 and 5,000 people were killed, and between 7,000 and 10,000 were injured.

6   Use of chemical weapons in the Syrian Civil War has been confirmed by the United Nations. Attacks took place near Damascus (2013), and near Aleppo (2013). A U.N. mission found likely use of the nerve agent Sarin and that the perpetrators likely had access to chemicals from the Syrian Army's stockpile. In August 2016, a report by the United Nations and the OPCW explicitly blamed the Syrian military of Bashar al-Assad for dropping chemical weapons (chlorine bombs) on the towns of Talmenes in April 2014 and Sarin in March 2015 and ISIS for using sulfur mustard on the town of Marea in August 2015. In April 2017, a chemical attack on Khan Shaykhun drew international condemnation and provoked the first U.S. military action against the Syrian government-controlled airbase at Shayrat.

7   Indian Chemical Export Controls System and the Australia Group, Rajiv Nayan, CBW Magazine, http://www.idsa.in/cbwmagazine/ IndianChemicalExportControlsSystemand%20theAustraliaGroup_ RajivNayan accessed 16-11-2015

# Chapter 6

# Biotechnology – The Key to Life

*"We didn't stay in the caves. We haven't stayed on the planet. With biotechnology, gene sequencing, we are not going to even stay within the limitations of biology"*

– *Jason Silva*

Mankind's advances in biological sciences can be traced back to Aristotle and Galen, during the Graeco-Roman period[1], followed by Avicenna in the Middle Ages. In the early modern period, European scientists were in the forefront of advances. In the 18th and 19th centuries, Darwin's theory of evolution and the germ theory of disease were major advances. In the 20 th century, the rediscovery of Mendel's work (of the previous century) on heredity led to advances in genetics, and led to the identification of Deoxyribo Nucleic Acid (DNA) as the basis for heredity. The deciphering of the structure of DNA by Watson and Crick in 1953 ignited a revolution in biology, leading to the understanding that there was a common basis at the molecular level for all life forms. The science of molecular biology and its applied field of biotechnology were transformed.

## The DNA revolution in life sciences

Biotechnology in its aspect of producing useful products from life forms had been known for long. Alcohol, Yoghurt and Cheese are some such products of traditional biotechnology known since ancient times[2]. But the discovery of DNA and its role in cellular processes revolutionized biotechnology. Till then classical biotechnology had to depend on natural processes for bringing about genetic changes which were usually random in character, and then selecting out the desired changed life forms, for example in the case of plant breeding. Now mankind had in theory possibility of deliberately making changes in the DNA of life forms, transferring genetic material across species, and even creating artificial genetic material. This marked the advent of modern biotechnology.

## Biotechnology development

But the development of these ideas in practice required considerable research and effort over decades. By 1975, techniques such as recombinant DNA had been discovered for modifying DNA and transferring genetic material across species. At the same time, techniques for sequencing DNA, or elaborating the precise structure advanced rapidly. There was a race among industry to use this technology to create useful products such as hormones and insulin. Genentech (USA) won the race to produce insulin using this new technology in 1978. The biotechnology industry grew rapidly into a promising real industry. In 1988, only five proteins from genetically engineered cells had been approved as drugs by the United States Food and Drug Administration (FDA), but this number jumped to over 125 by the end of the 1990s. However, the functioning of DNA and the expression of genes especially in higher organisms turned out to be far more complex than anticipated. New technology such as the polymerase chain reaction, has enabled rapid genetic analysis, including sequencing entire genomes of life forms. Biotechnology has numerous applications and great potential for transforming agriculture, animal and human health, industry, environment, and energy, through use of genetically modified organisms. Recently, a new gene-editing technology known as CRISPR (Clustered Regularly Interspaced Palindromic Repeats)[3] has been called "the biggest biotech discovery of the century." It is a quick, easy and effective way to edit the genes of any species, including humans in weeks rather than months and years. It could spark a whole new wave of innovation in biotechnology and has potential applications for the ability to create new biofuels, materials, drugs and foods within much shorter time frames at a relatively low cost. At the same time it raises numerous regulatory issues[4] that need to be considered.

## International Cooperation in Biotechnology

Developing countries saw the emerging field of biotechnology as a great opportunity to transform their agriculture and health. However, they lacked the technology and knowhow which was largely available in the developed countries. In 1981 the United Nations Industrial Development Organization (UNIDO) located in Vienna, Austria, convened a panel of distinguished scientists to consider the issue of transferring biotechnology know-how to the developing world[5]. The result was a proposal to establish the International Centre for Genetic Engineering and Biotechnology (ICGEB) as a centre of excellence. In 1982 the concept was approved by a ministerial-level conference of developed and developing nations and the first steps to establish the Centre were taken in 1983 when the Statutes of ICGEB were signed by 26 countries.

A Panel of Scientific Advisors together with a Preparatory Committee, consisting of representatives from ICGEB Member Countries, negotiated the establishment and subsequent development of ICGEB under the aegis of UNIDO. Six countries (Belgium, Cuba, India, Italy, Pakistan, and Thailand made offers to host the Centre. The joint proposal of Italy and India was finally chosen, after a rather divisive voting process[6] at a Plenipotentiary Meeting on the Establishment of the ICGEB held on 3-4 April 1984, which decided that the Centre will be located in two components, one in Trieste, Italy, and the other in New Delhi, India. The decision took into account the offers made by the governments of the two countries which, among other things, specified general areas of concentration between the two components of the ICGEB, namely industrial microbiology in the Trieste Component and agriculture, human, and animal health in the New Delhi Component[7].

An interim programme was launched that ran from 1986 to 1989 during which the ICGEB established two component laboratories, one in Trieste (Italy) and the other in New Delhi (India)[8]. In 1989 a Five Year Programme with a budget of US$56 million was initiated for the period, 1989 to 1994. In 1994, the ICGEB became an independent intergovernmental organization and in 1995 it became a UN agency. It currently has 64 members, which are mostly developing countries and a further 22 countries are considering accession to or ratification of the statutes. The main functions of the ICGEB are: to transfer knowledge in genetic engineering and biotechnologies to emerging and developing countries, and to supervise its aims and use; to carry out research activities and training and specialized courses with long-term study grants.

The ICGEB was financed by the Italian government (Euro 12.4 million euro annually, about 85% of the ICGEB budget) and, to a lesser extent by India, and since 1999 it also collects the obligatory contributions of its 86 member countries. In 2005 the ICGEB Board of Governors approved establishment of its third component in Africa, and in November 2006 it ratified the choice of South Africa as the country to host it. The third component of the ICGEB was inaugurated in Cape Town, South Africa in September 2007. It also has a network of 40 Affiliated Centres (national laboratories located in developing countries). In recent years it has suffered from financing difficulties. Since 2014, Italy has stopped financing for the ICGEB[9] and the three components located in New Delhi, Trieste and Cape Town are to be funded exclusively by the respective host countries. This transition is underway and efforts are being made to maintain the level of ICGEB's activities despite reduction in funding.

The ICGEB has been partially successful in stimulating transfer of knowledge in biotechnology to developing countries. But membership has been confined largely to developing countries and economies in transition. The advanced countries, with the exception of Italy, have not joined. On the other hand, several renowned scientists from institutions in advanced countries are members of its 15 member Scientific Council. ICGEB also receives additional funding for research projects through grants from government agencies, charities, non-profit foundations and the private sector, including from the European Commission, and institutions in the UK, USA, Canada, France, etc.

Has the ICGEB been a success? In terms of technology transfer it has concluded over 80 agreements with industrial partners in Argentina, Brazil, China, Cuba, Egypt, UAE, India, Iran, Pakistan, South Africa, Sri Lanka, Turkey, Uruguay, USA (a non member state) and Venezuela, predominantly to produce pharmaceuticals obtained through genetic engineering. It has produced over 2700 publications in highly acclaimed scientific journals, provided over 900 fellowships of 3-4 years' duration to young scientists (awarded since 1988), involved over 1200 participants in meetings and courses each year, carried out some 500 research projects funded in developing countries (for over 19 million Euro) and has been granted over 60 patents. The ICGEB has faced financial difficulties due to the reduction of financing from Italy and India due to various reasons, which has not been compensated by contributions from other member states. As a case of an international centre set up with the vision of providing access to cutting edge technology, its success has been limited, largely due to lack of political commitment from governments of advanced countries, barring a few.

## The Human Genome Project

The Human Genome Project was launched in 1990 in the US as a massive international collaborative effort to map and sequence all the genes of the species of man Homo Sapiens[10]. The United States Department of Energy, seeking data on protecting the genome from the mutagenic (gene-mutating) effects of radiation, became involved in 1986, and established an early genome project in 1987. The National Institute of Health was added in 1988, and in 1990 a five year research plan (of a 15 year projected $ 3 billion effort) was launched In addition to the United States, the international consortium comprised institutions in the United Kingdom, France, Australia, China and myriad other spontaneous relationships. The National Center for Human Genome Research (NCHGR) was set up as the coordinating centre. Further research plans for 1993-98 and 1998-2003 followed. By 2001 the project had published 90 percent of the genome's three billion base-pairs. The essentially complete genome sequencing was completed on April 14, 2003, two years earlier than planned. A parallel project was conducted outside of government by Celera Genomics, which was formally launched in 1998. Most of the government-sponsored sequencing was performed in twenty universities and research centers in the United States, the United Kingdom, Japan, France, Germany, and China.

The HGP project is an example of successful international scientific collaboration. Its success can be attributed to several factors. The strong political commitment by the most advanced country, the US, backed by long term financing and institutional support. Around this core, it was possible to build collaborations extending to other governments, and the private sector. In the research effort, Celera Genomics (led by Dr. Craig Venter) provided scientific breakthroughs that made much faster progress possible. These are important lessons to be learnt for such projects in future.

## Regulatory issues in Biotechnology

In a fast advancing and cross cutting technology like biotechnology, it is inevitable that regulatory systems have to firstly be devised and adapted to change. Biotechnology has promising applications in human and animal health, agriculture, industry and environment, for example. Regulatory systems were initially focused on protecting health and the environment. As the US was the world leader in biotechnology, it is useful to examine the evolution of its regulatory effort. As in any other technology which has wide application, biotechnology products and services are found in many economic sectors, subject to different regulatory regimes and institutions. It was felt that a separate legislation on biotechnology would quickly become outdated and could hamper innovation. The same could happen if another wide ranging technology came into existence. Therefore in 1986, the US government went in for a coordinated regulatory framework for biotechnology to coordinate the work of various regulatory regimes. It noted that "existing statutes provide a basic network of agency jurisdiction over both research and products, assuring reasonable safeguards for the public and the environment"[11].

Biotechnology involves the modification of living organisms and the use of them to produce useful products. Therefore the release of genetically modified organisms into the environment could pose a threat to the ecosystem, especially in large quantities or if the released life form is capable of independent existence and reproduction. In agriculture, genetically modified crops pose such a problem, despite the better yields and quality of the crops. The impact on the ecosystem is difficult to determine and controlled field trials are required. The international trade involving these crops across national boundaries requires a harmonized international approach. The issue of informed choice by consumers is also important and may require special labeling. The Convention on Biodiversity seeks to achieve this. Research and development, and applications of biotechnology require special measures to prevent accidental release of genetically modified organisms into the environment. There is special concern when research involving modified pathogens is involved. Thus biotechnology brings with it a host of regulatory problems, cutting across various economic sectors, and impacting international trade as well.

## Release of and Transboundary movement of GMOs

Regulations regarding the release of genetically modified organisms (GMOs) outside the laboratory vary widely by country. Countries such as the United States, Canada, Lebanon and Egypt are more liberal when assessing safety, while the European Union, Brazil and

China authorize GMO cultivation on a case-by-case basis. Many countries allow the import of GM food with authorization, but either do not allow its cultivation, or have provisions for cultivation, but do not produce any GM products. Most countries that do not allow for GMO cultivation do permit research activities on GMOs.

Labeling requirements for GMO products is required in 64 countries. Labeling can be mandatory beyond a threshold GM content level or voluntary. There is general scientific agreement that food derived from GM crops poses no greater risk than conventional food. There is no evidence to support the idea that the consumption of approved GM food has a detrimental effect on human health. Some scientists and advocacy groups, such as Greenpeace and World Wildlife Fund, have however called for additional and more rigorous testing for GM foods.

Special problems may arise where genetically modified organisms are approved for use in one country but not in another. The GMOs may leak out and contaminate other non-GMO products that are being exchanged. Hence stringent testing may be necessary to detect such leakages and control them. The EU for example does not allow food products containing unapproved GMOs, and requires labeling if the GMO content exceeds 0.9 percent. Exporting countries may have to ensure that there is no contamination from unapproved GMOs. The European Commission has noted[12] that "the increase in the number of countries growing GM crops, as well as increases in the type of GM crops .are likely to result in an increasing number of applications for authorization in the EU, which could lead to an increasing number of issues with low level presence (LLP) of unauthorized GMOs in imported products (feed, food and seeds)."

## Convention on Biodiversity (CBD)

The Convention on Biological Diversity[13] was negotiated under the auspices of the United Nations Environment Programme (UNEP). It entered into force on 29 December 1993. 196 countries and territories have become Parties. The three goals of the CBD are to promote the conservation of biodiversity, the sustainable use of its components, and the fair and equitable sharing of benefits arising out of the utilization of genetic resources. The CBD Secretariat is located in Montréal, Canada. The Subsidiary Body on Scientific, Technical and Technological Advice (SBSTTA), which advises the Conference of the Parties (COP), meets several months prior to each COP.

Biotechnology has enabled scientists to genetically and biochemically modify plants, animals and micro-organisms to create genetically modified organisms (GMOs). Many countries with biotechnology industries already have domestic legislation in place intended to ensure the safe transfer, handling, use and disposal of GMOs and their products. These precautionary practices are collectively known as "biosafety." However, there are no binding international agreements addressing situations where GMOs cross national borders. Article 19 of the CBD provides for Parties to consider the need for and modalities of a protocol on biosafety.

The Cartagena Protocol on Biosafety to the Convention on Biological Diversity is an international treaty governing the movements of genetically modified organisms (GMOs) resulting from modern biotechnology from one country to another. It was adopted in 2000 as a supplementary agreement to the Convention on Biological Diversity and entered into force in September 2003. The Protocol seeks to protect biological diversity from the potential risks posed by GMOs and establishes an advance informed agreement (AIA) procedure for ensuring that countries are provided with the information necessary to make informed decisions before agreeing to the import of such organisms into their territory. The Protocol also establishes a Biosafety Clearing-House to facilitate the exchange of information on GMOs and to assist countries in the implementation of the Protocol. So far 170 Parties have signed the Protocol.

The Nagoya Protocol on Access and Benefits Sharing (ABS) to the Convention on Biological Diversity is a supplementary agreement to the CBD. It provides a transparent legal framework for the effective implementation of one of the three objectives of the CBD: the fair and equitable sharing of benefits arising out of the utilization of genetic resources. It was adopted on 29 October 2010 in Nagoya, Japan and entered into force in October 2014, and now has 68 Parties.

The Nagoya Protocol addresses traditional knowledge associated with genetic resources with provisions on access, benefit-sharing and compliance. It also addresses genetic resources where indigenous and local communities have the established right to grant access to them. Contracting Parties are to take measures to ensure these communities' prior informed consent, and fair and equitable benefit-sharing, keeping in mind community laws and procedures as well as customary use and exchange.

## Biological Weapons Convention

In the past, biological toxins or infectious agents such as bacteria, viruses, and fungi were used with the intent to kill or incapacitate humans, animals or plants as an act of war. Biological weapons are living organisms or replicating entities (viruses) that reproduce or replicate within their host victims. Entomological (insect) warfare is also considered a type of biological weapon.

The Geneva Protocol of 1925 prohibited the use of chemical weapons and biological weapons. During World War II, the United Kingdom established a BW program at Porton Down and weaponized tularemia, anthrax, brucellosis, and botulism toxins and brought them into industrial production. The USA set up a large industrial complex at Fort Derrick, Maryland in 1942 for biological and chemical weapons development and the mass production of anthrax spores, brucellosis, and botulism toxins, although the war was over before these weapons could be of much operational use. The Imperial Japanese Army did research on BW, conducted experiments on prisoners, and produced biological weapons for combat use. Biological weapons were used against both Chinese soldiers and civilians in several military campaigns. In Britain, the

1950s saw the weaponization of plague, brucellosis, tularemia, equine encephalomyelitis and vaccinia viruses, but the programme was unilaterally cancelled in 1956. The United States Army Biological Warfare Laboratories weaponized anthrax, tularemia, brucellosis, Q-fever and others. In 1969, the UK and the Warsaw Pact, separately, introduced proposals to the UN to ban biological weapons, and US terminated production of biological weapons, allowing only scientific research for defensive measures. Biological weapons were generally not militarily effective though they could hamper military operations. Negotiations started on a Convention to ban biological weapons.

The Biological Weapons Convention (BWC) is a legally binding treaty[14] that outlaws biological arms. After being discussed and negotiated in the United Nations' disarmament forums starting in 1969, the BWC opened for signature on April 10, 1972, and entered into force on March 26, 1975. It currently has 174 states-parties and 12 signatory states. It was the first multilateral disarmament treaty banning the production of an entire category of weapons. The BWC bans (1) the development, stockpiling, acquisition, retention, and production of Biological agents and toxins of types and in quantities that have no justification for prophylactic, protective or other peaceful purposes; (2) Weapons, equipment, and delivery vehicles designed to use such agents or toxins for hostile purposes or in armed conflict; and (3) the transfer of or assistance with acquiring the agents, toxins, weapons, equipment, and delivery vehicles described above. The convention further requires states-parties to destroy or divert to peaceful purposes the agents, toxins, weapons, equipment, and means of delivery described above within nine months of the convention's entry into force.

The BWC does not ban the use of biological and toxin weapons but reaffirms the 1925 Geneva Protocol, which prohibits such use. It also does not ban biodefense programs. The absence of any formal verification regime to monitor compliance has limited the effectiveness of the Convention. Negotiations towards an internationally binding verification protocol to the BWC took place between 1995 and 2001 but in July 2001, the US administration, after conducting a review decided that the proposed protocol did not suit the national interests of the United States. The US has been opposed to verification provisions probably due to the desire to protect its biotech industry from intrusive international inspections and avoid leakage of technology and knowhow in a field in which it has been the leader. In this situation, a mechanism of reporting by parties of information through confidence building measures (CBM)s has served to strengthen implementation of the Convention

The BWC is reviewed every five years, the last being held in 2011. The Final Declarations of the Review Conferences contain additional commitments and interpretations of the BWC provisions. The next review is due in November 2016 and will provide an opportunity to deal with new developments including bioterrorism (see below)

## Potential of new Biological Weapons

Many pathogens found in nature have been the subject of research on biological weapons. However, biotechnology has made it possible to genetically modify pathogens to radically alter their characteristics. This has raised the possibility of research and development of more deadly biological agents not found in nature. New and emerging pathogens such as H1N1, Avian Flu, Ebola, etc could be modified to make them more dangerous. Such pathogens could be highly destructive not only for humans, but for animals and food crops. This has led to discussions on the need to regulate scientific research[15] into such sensitive areas, or at least restrict the publications of such research.

## Bioterrorism

In recent times, non-state actors such as terrorist groups have attempted to gain access to weapons of mass destruction including biological weapons. While such weapons have relatively little use against military forces, they can cause casualties if used against civilian populations. There have been cases of use of anthrax[16] and ricin[17], a toxin found in castor seed. Among some of the advantages of bioweapons are (1) Relatively low cost in developing them (2) Easy availability of materials and technology (3) Ease of multiplying and manufacturing biological agents (4) Ease of deploying them especially in large population concentrations. If a medium corporation can carry out research and development of biotechnology products, then terrorist groups could also do so. The production of harmful biological agents may be a constraint, but natural mechanisms exist such as use of pigs as biological mixing vehicles[18] to produce new agents from known ones. Such scenarios may seem remote, but cannot be ruled out. Countering bioterrorism requires early detection and rapid response and therefore close cooperation between public health authorities and law enforcement and at the international level. However, such cooperation is currently lacking. Also, international control and monitoring regimes on sensitive biotechnology materials, research and technology may become necessary[19]. The current approach to prevent the proliferation of biological weapons development relies heavily on the existence of effective export controls for equipment particularly suitable for use in the production or dissemination of biological agents. The Australia Group (AG), an informal forum that brings together 41 countries and the European Union, seeks to harmonize national export controls on such equipment.

Governments have responded to the threat of bioterrorism by the creation of lists of controlled materials and equipment and restricting access to them and creating legal tools for the prosecution of individuals on the basis of possession. In an age of terrorism biological weapons are perfectly suited for asymmetric warfare where the relatively low costs of producing such weapons combined with their potential for amplification through communicability have a disproportionately strong effect on targeted populations. Consequently, biological weapons are likely to remain very attractive to terrorists[20]. New technologies are likely to emerge in future that will radically transform biological warfare offensive and defensive possibilities. Even without envisioning new biological agents, using synthetic biology, the technology already exists for significantly enhancing the lethality of biological weapons.

## Intellectual Property issues

Biotechnology has raised certain intellectual property issues. In the pharmaceutical industry, useful molecules which are discovered can be patented. This product patent regime introduced post 2005 effectively gives the patentee an exclusive right to the molecule. Patent applications have been made for patenting species such as Neem (Azadirachta Indica)[21] and Turmeric (Curcuma Longa)[22] , which have medicinal properties. These have been contested on the grounds that traditional knowledge cannot be patented. But can genes be patented ? The situation is complex and there are conflicting positions. Genes exist in nature, so they cannot be claimed to be discovered, unless one speaks of synthetic gens not found in nature. However, the practical use gene in a particular situation may be regarded as new knowledge and patent rights could be considered for it. The situation is evolving.

## Ethical issues at the frontier of biotechnology

A number of ethical issues have arisen in the field of biotechnology, especially in human health. Protection of persons participating in gene therapy trials is a concern. Genetic sequencing makes is increasingly likely that compromising information about a person's future health is going to become available. Privacy issues arise. Ethical issues may arise when research involves embryos, fetal tissue, cloning, or other controversial questions. Should it be permitted to alter human genes to produce offspring of a particular sex or other desired characteristics such as physical or mental ability? Stem cell research promises great good and is a worthy scientific priority. Obtaining stem cells from people without seriously harming people in the process can be ethical. However, obtaining stem cells from human embryos may not be regarded as ethical because it necessarily involves destroying those embryos. The recent success in genomic editing of human embryos for the first time as reported in Aug 2017, by a US-South Korean joint team to remove defects in a gene that are implicated in hypertrophic cardiomyopathy opens up new clinical possibilities but also raises a myriad of ethical and legal issues.

Some broader questions may need to be addressed[23]. Should research be limited and, if so, how should the limits be decided? How should the limits be enforced nationally and internationally? Are there fundamental issues with creating new species? Will transgenic interventions in humans create physical or behavioral traits that may or may not be readily distinguished from what is usually perceived to be "human"? On many of these ethical issues consensus has still to emerge even at the national level.

## India's approach

In 1986, India set up Department of Biotechnology (DBT)[24] under the Ministry of Science and Technology responsible for promoting development and commercialization in the field of modern biology and biotechnology in India. The DBT has promoted the growth and application of biotechnology in the broad areas of agriculture, health care, animal sciences, environment, and industry, patenting of innovations, technology transfer to industries. Other priorities include molecular biology of human genetic disorders, plant genome research, development, food biotechnology, and setting up of micropropagation parks. In December 2015, the DBT launched the National Biotechnology Development Strategy 2015-2020 programme to intensify research. The mission is backed with significant investments to create new products, and creating a strong infrastructure for research and development. The Indian biotech industry holds about 2 per cent share of the global biotech industry. The biotechnology industry in India, comprising about 800 companies, is growing at an average rate of about 20 per cent[25]. The Indian biotechnology sector is expected to grow from the current US$ 5-7 billion to US$ 100 billion by 2025.

India has a Biosafety Research programme to facilitate the implementation of biosafety procedures, rules and guidelines under Environment (Protection) Act 1986 and Rules 1989 to ensure safety from the use of Genetically Modified Organisms (GMOs) and products thereof in research and application to the users as well as to the environment. A three tier mechanism comprising Institutional Biosafety Committees (IBSC) at the Institute/ company level; the Review Committee on Genetic Manipulation (RCGM) in the Department of Biotechnology; and the Genetic Engineering Approval Committee (GEAC) in the Ministry of Environment & Forests (MoE&F) for granting approval for research and development activities on recombinant DNA products, environmental release of genetically engineered (GE)crops and monitoring and evaluation of research activities involving recombinant DNA technology has been established. Indian GMO Research Information System (IGMORIS) is a database on activities involving the use of GMOs and products in India to make available objective and realistic scientific information relating to GMOs and products under research and commercial use to all stakeholders including scientists, regulators, industry and the public in general.

In order to update and streamline the regulatory system for the biotechnology sector, in 2013 the Government introduced a Bill to sets up an independent authority, the Biotechnology Regulatory Authority of India (BRAI)[26], to regulate organisms and products of modern biotechnology. BRAI will regulate the research, transport, import, containment, environmental release, manufacture, and use of biotechnology products. Regulatory approval by BRAI will be granted through a multi-level process of assessment undertaken by scientific experts. However the Bill could not be passed and has lapsed. The biotechnology industry has advocated the need for an efficient regulatory framework that can decide on approvals quickly and allow for growth of the industry. On the other hand, anti-GMO activists have called for a cautious approach and for greater attention to biosafety, biodiversity, and impact on traditional farming.

## Outlook

Biotechnology is likely to witness rapid advances in future, and this is likely to give rise to numerous issues needing action at the international level. Examples are GMOs and their impact in biodiversity and on commercial agriculture, synthetic genomes and their impact on biodiversity, IPR related issues, etc. Such issues will need to be handled through diplomacy. The caution advocated by some civil society groups (for example on the introduction of GMO foods crops) is justified and should not be dismissed as anti-technology, and the issues raised need to be addressed objectively on the basis of scientific evidence. It must also be kept in mind that society needs to absorb and understand the full implications of new technological developments before they gain wide acceptance, and for this process transparency and effective communication of scientific facts is essential.

## Endnotes

1 The Origins Of Biology, http://science.jrank.org/pages/8467/Biology-Origins-Biology.html accessed 29-1-2016

2 Biotechnology in the realm of History, A.S. Verma et al, 2011, http://www.ncbi.nlm.nih.gov/pmc/articles/PMC3178936 accessed 29-1-2016

3 Everything you need to know about why CRISPR is such a hot technology, Washington Post, 4 November 2015 https://www.washingtonpost.com/news/innovations/wp/2015/11/04/everything-you-need-to-know-about-why-crispr-is-such-a-hot-technology accessed 29-1-2016

4 How should the applications of genome editing be assessed and regulated? Robin Fears, eLife, 4 Apr 2017, https://www.ncbi.nlm.nih.gov/pmc/articles/PMC5380431/ , accessed 13-4-2017

5   Biotech for development - Indico, M Giacca, http://indico.ictp.it/event/
    a13223/session/26/contribution/69/material/slides/0.pdf accessed 29-
    1-2016

6   Unido's ICGEB: A Case of Petty Politics Prevailing Over Scientific
    Peer-Reviews, P Matangkasombut, 1984, https://www.researchgate.net/
    publication/16598037_Unido's_ICGEB_a_case_of_petty_politics_
    prevailing_over_scientific_peer-reviews accessed 29-1-2016

7   Meeting of Panel of Scientific Advisors of the ICGEB June 1985,
    https://profiles.nlm.nih.gov/ps/access/BBGMNQ.pdf   accessed 29-1-
    2016

8   ICGEB website, http://www.icgeb.org/about-the-centre.html accessed
    29-1-2016

9   I am sure ICGEB will do better under DBT, Rahul Koul, BioSpectrum,
    23   April   2014,   http://www.biospectrumindia.com/biospecindia/
    interviews/213502/-i-icgeb-dbt , accessed 27-1-2017

10 All about the Human Genome Project, http://www.genome.
    gov/10001772 accessed 27-1-2016

11 Federal Register, 93-4266, 26 Feb 2002, Office of Science and
    Technology Policy, Exercise of Federal Oversight Within Scope of
    Statutory Authority: Planned Introductions of Biotechnology Products
    Into the Environment.

12 GMOs: EU's legislation on the right track, evaluation reports conclude,
    Press Release, 28 October 2011, http://europa.eu/rapid/press-release_
    IP-11-1285_en.htm?locale=en accessed 27-1-2016

13 Convention on Biological Diversity website, https://www.cbd.int  ,
    accessed 27-1-2016

14 The Biological Weapons Convention (BWC) At A Glance, Arms
    Control   Association,   https://www.armscontrol.org/factsheets/bwc
    accessed 27-1-2016

15 Scientists call for curbs on own research on deadly bird flu virus, The Guardian, 3 Feb 2012, https://www.theguardian.com/world/2012/feb/03/bird-flu-virus-scientists-warning , accessed 13-4-2017

16 Bacillus anthracis Bioterrorism Incident, Kameido, Tokyo, 1993, CDC, http://wwwnc.cdc.gov/eid/article/10/1/03-0238_article accessed 29-1-2016

17 Intel Hints at Al-Qaida Steps Toward Ricin Strikes, NTI, 2013, http://www.nti.org/gsn/article/intel-hints-al-qaida-steps-toward-ricin-strikes accessed 29-1-2016

18 India swine flu has mutated to become more deadly as virus claims 1,200 lives, International Business Times, 11 March 2015, http://www.ibtimes.co.uk/india-swine-flu-has-mutated-become-more-deadly-virus-claims-1200-lives-1491498 accessed 29-1-2016

19 Biotechnology E-commerce: A Disruptive Challenge to Biological Arms Control, May 2015, http://www.nonproliferation.org/wp-content/uploads/2015/05/biotech_ecommerce.pdf accessed 29-1-2016

20 The future of biological warfare, A. Casadevall, Microbial Biotechnology 5, 584-7, 2012, https://www.ncbi.nlm.nih.gov/pmc/articles/PMC3815869/pdf/mbt0005-0584.pdf , accessed 13-4-2017

21 India wins landmark Neem patent battle in Europe, infochange India, 9 March 2005, http://infochangeindia.org/trade-a-development/news-scan/india-wins-landmark-neem-patent-battle-in-europe.html accessed 27-1-2016

22 India foils US firm bid to patent turmeric, Hindustan Times, 28 January 2013, http://www.hindustantimes.com/delhi/india-foils-us-firm-bid-to-patent-turmeric/story-kjGGebkCYbLAaaosNeR1XJ.html , accessed 27-1-2016

23 Ethical Issues in Genetic Engineering and Transgenics, Linda MacDonald Glenn, Action bioscience, Nov 2013, http://www.actionbioscience.org/biotechnology/glenn.html , accessed 27-1-2016

24 Department of Biotechnology website, http://www.dbtindia.nic.in accessed 27-1-2016

25 Biotechnology industry in India, IBEF, http://www.ibef.org/industry/biotechnology-india.aspx accessed 27-1-2016

26 The Biotechnology Regulatory Authority of India Bill, 2013 Legislative Brief, PRS Legislative Research, http://www.prsindia.org/uploads/media/Biotech%20Regulatory/Brief-%20BRAI%20Bill%202013.pdf, accessed 27-1-2016

# Chapter 7

# Information and Communication Technology – Instant Information

*"This is just the beginning, the beginning of understanding that cyberspace has no limits, no boundaries. ..National law has no place in cyberlaw. Where is cyberspace? ... Cyberlaw is global law, which is not going to be easy to handle, since we seemingly cannot even agree on world trade of automobile parts."*

– *Nicholas Negroponte*

## The ICT Revolution

The growth of information and communications technology (ICT) can be traced back to the development of the telegraph services (1839), the telephone (1871), radio transmission (1900), and television broadcasting (1928). This was followed by the computer (1945), and microprocessors (1970), communications satellites (1962), mobile cellular services (1979), and the availability of the internet and World Wide Web (1990)[1]. These mutually reinforcing developments led to a massive development of ICT since the 1970s that has transformed almost every aspect of human activity, the so-called ICT revolution. Some of the important distinct features of this revolution are - (1) rapid growth in information processing capacity and sharp drop in costs (2) the growth of decentralized and distributed information and communication capacity across populations through increasingly mobile platforms (3) the growth of transnational communications and social networks, and information services easily accessible to the population at low or no cost (4) the sharp reduction in costs, increase in reliability, capacity and speed of communications. These developments have transformed almost every aspect of human life, even in the less developed countries, and has brought world much closer together, through globalization of ICT. The social, economic, security and political consequences of these changes are profound. In international relations, there are important consequences in security, military, political, economic and social domains.

## Indian ICT development

The advent of the ICT revolution offered India a great opportunity. India's manufacturing sector faced handicaps of inadequate logistics, infrastructure, and energy supply, as well as environmental constraints, poor regulatory systems and outdated technology and business processes, and protectionist regulations. In contrast, in the ICT sector, these difficulties did not exist. ICT firms could be set up with relatively small investment, low energy and physical logistics requirements, and could benefit from the large pool of English knowing human resources. There was relatively little government interference, and the government in the post-reform period adopted many promotional measures. The one requirement was of good and low cost telecommunications and data linkages with the outside world, and this could be achieved through telecom sector reforms initially in the large metros and later on in second and third tier cities. Low cost human resources for software work enabled India to benefit from the global needs such as global demand for Y2K adjustments of 1999-2000. The ICT sector grew rapidly, in an atmosphere free of government interference.

The Indian government, especially in the post reform period after 1991, realized the great potential in the Indian ICT industry. A number of promotional measures were taken[2]. In May 1998, India set up a high level National Task Force on Information Technology and

Software Development to draw up a national policy on informatics aimed at helping India become an IT superpower within 10 years. This followed the 1994 reforms opening up the telecom sector to the private sector. The Task force recommended an overhaul of the telecom policy to foster rapid IT sector development. Consequently a New Telecom Policy was announced in March 1999. Within 90 days of its establishment, the Task Force produced an extensive background report on the state of technology in India and an IT Action Plan with 108 recommendations. The Task Force could act quickly because it built upon the experience and frustrations of state governments, central government agencies, universities, and the software industry. The Information Technology Act 2000 created legal procedures for electronic transactions and e-commerce and laid the foundation for further growth of the IT industry.

Another important factor was the strong partnership between Indian IT firms and the Indian diaspora, especially in the US. In the 1990s, Indian professionals entered the United States in increasing numbers, reaching 1.7 million by 2000. This immigration consisted largely of highly educated technologically proficient workers and contributed to US business growth in high tech sectors.

The growth of the knowledge economy facilitated the growth of an entrepreneurial class of immigrant Indians, which helped aid in promoting technology-driven growth. A large pool of highly qualified IT entrepreneurs and professionals were available in the US. Partnerships developed to exploit the advantages offered by both sides – from India low cost skilled manpower, and from the US cutting edge IT and market access. The success of Information Technology in India not only had economic repercussions but also had significant political consequences. India's reputation both as a source and a destination for skilled workforce helped it improve its relations with a number of world economies.

The Indian Information Technology and enabled services (IT-ITES) industry has grown substantially. Its revenue is estimated at USD 119.1 billion in FY 2014-15 as compare to USD 76.3 billion in FY 2010-11, registering an average annual growth rate of around 11%. Exports in this sector grew from USD 59 billion in 2010-11 to USD 98.1 billion in 2014-15 representing an average annual growth rate of 13%. In 2013, India accounted for 3% of global service exports in with sixth rank, while it accounts for 2.8% of global services imports in with ninth rank.

## IT and International Service Trade Issues.

The growth of IT based services exchanges across national boundaries raised several issues not found in the case of merchandise trade. IT services could be provided from the exporting country directly to entities in the consuming country (Cross border supply or Mode 1)[3]. They could also be purchased by entities from the consuming country in the exporting country (Consumption abroad or Mode 2). Another way would be for the exporting country to send personnel to the consuming country for providing the services (presence of natural persons, or Mode 4). Finally, entities in the exporting country could establish a presence in the consuming country and provide services to consumers (commercial presence, or Mode 3).

Given India's comparative advantages, in the IT sector the most important modes were Mode 1 and Mode 4 where business regimes in the consuming country play a big role. Mode 1 services are rendered through better data and communications linkages. This includes off-shoring of work for foreign entities. Mode 4 involved movement of persons to the consuming country which raised questions of visas and work permits. Both Modes 1 and 4 result in work being transferred to India with consequent loss in jobs and value addition in the consuming country. This provoked opposition from groups within the consuming country who could be affected adversely. On the other hand, business enterprises see this as a means of bringing down costs and improving competitiveness, and consequently greater economic output. India's main interests are for liberalization of services under Modes 1 and 4 of General Agreement on Trade in Services (GATS). It is seeking recognition as a "data-secure" nation for improving market access for Indian information technology companies. In addition, India would like a more liberal visa regime for its professionals.

The General Agreement on Trade in Services (GATS) under the WTO, which entered into force in 1995[4], seeks to liberalize the regime for trade in services, including IT services. Countries are encouraged to make commitments in regard to market access for services. 22 plurilateral groups have been formed at the WTO in service sectors/areas. India has received plurilateral requests for greater liberalization in 14 different services sectors, including Telecom, Finance, Maritime, Environment, Education, and Air transport, Energy, Audio Visual and Retail[5]. India is the coordinator of the plurilateral requests on Mode 1 (cross border supply) and Mode 4 (Movement of Natural Persons) - the core areas of its interest in the services negotiations - India is also a cosponsor of plurilateral requests on Computer and Related Services (CRS) and Architectural, Engineering and Integrated Engineering Services. Developed countries, particularly the US and the EU have not been forthcoming in offering substantial openings in Mode 4. Even where these commitments have been offered, they lack in sectoral spread.

The lack of progress at the GATS has resulted in the launch of a parallel effort in 2013 to negotiate a plurilateral agreement called the Trade in Services Agreement (TISA)[6] by a group of 25 WTO members, including the 28 member European Union. Together they account for 70% of world services trade. The architecture of TISA presents a key divergence from the GATS. While the GATS allowed countries to choose specific sectors and modes for national treatment and market access, the TISA expects countries to liberalize services in essentially all modes and sectors. Negotiations for a plurilateral Trans Pacific Partnership (TPP), and a Transatlantic Investment and Trade Partnership (T-TIP) could also impact services trade regimes.

In parallel with efforts in the WTO, India has sought to secure more liberal services trade regimes for its exports through bilateral and regional trading arrangements. Examples are the Comprehensive

Economic Partnership/Cooperation Agreements (CEP/CA) with Malaysia, Singapore, Republic of Korea, and Japan[7].

## Cyber Security Issues

Rapid development of ICT has also brought unprecedented threats with them[8]. Cyber security – defined as the protection of systems, networks and data in cyberspace has become a critical issue. These threats include cyber crime, cyber war, and cyber terrorism. Cyber crime is conducted by individuals working alone, or in organized groups, intent on extracting money, data or causing disruption. Cyber war involves nation state conducting sabotage and espionage against another nation in order to cause disruption or to extract data. Cyber terrorism involves an organization, working independently of a nation state, conducting terrorist activities through the medium of cyberspace. These threats are considered serious- for example, the UK's National Security Strategy identifies cyber attack as one of the four highest-priority risks faced by the UK.

Cyberspace is particularly difficult to secure[9] due to a number of reasons: malicious actors can operate from anywhere in the world, the growing linkages between cyberspace and physical systems, and the difficulty of reducing vulnerabilities and consequences in complex cyber networks. Critical infrastructure is increasingly subject to cyber intrusions that pose new risks. Wide scale or high-consequence events could cause harm or disrupt services upon which the economy and the daily lives of people depend. Strengthening the security and resilience of cyberspace has become important.

International cooperation between governments and other entities is important in the context of the borderless and increasingly sophisticated nature of cyberthreats. Since 1998, when the issue was brought to the UN by Russia, the UN had established four Groups of Governmental Experts (GGE) from 2004 onwards[10] to consider this matter. A significant advance was achieved in 2013, when the group agreed on the centrality of the Charter of the United Nations and international law as well as the importance of States exercising responsibility. In 2015, the group's report focused on (1) existing and emerging threats; (2) norms, rules, and principles for the responsible behaviour of states; (3) confidence-building measures (CBMs); (4) international cooperation and capacity-building; (5) the applicability of international law, and (6) recommendations for future work. The report was approved by the General Assembly and will be taken forward by a new GGE in 2016-17. The GGE's work may appear slow, but given the complexity and technical nature of the issues, and the diversity in approaches and interests of the countries represented in the GGE, this is to be expected. One indication of differences is the parallel effort by the six countries of the Shanghai Cooperation Organization (SCO) to promote an International Code of Conduct for Information Security[11] in the UN.

The relatively homogenous region of Europe, meanwhile, witnessed the birth of the Convention on Cybercrime, also known as the Budapest Convention on Cybercrime or the Budapest Convention. This is the first international treaty seeking to address Internet and computer crime by harmonizing national laws, improving investigative techniques, and increasing cooperation among nations. It was drawn up by the Council of Europe in Strasbourg, France, with the active participation of the Council of Europe's observer states Canada, Japan, South Africa and the United States. The Convention was adopted on 8 November 2001 and by September 2015, 47 states have ratified the convention, while a further seven states had signed the convention but not ratified it. States outside Europe that have ratified the treaty are Australia, Canada, Dominican Republic, Japan, Mauritius, Panama, Sri Lanka, and the United States. However, many countries outside Europe as well as Russia are not parties, thus limiting the coverage of the Convention. The Convention aims principally at: harmonizing the domestic criminal law in the area of cyber-crime; providing for the investigation and prosecution of such offences; and setting up a fast and effective regime of international cooperation. The Convention has been supplemented by an Additional Protocol making any publication of racist and xenophobic propaganda via computer networks a criminal offence. Currently, cyber terrorism is also studied in the framework of the Convention.

The role of a global facilitator of cybersecurity cooperation was entrusted to the International Telecommunications Union (ITU) at the World Summit on the Information Society (WSIS) held in 2003 and 2005. The ITU launched an international cyber security forum in 2007 – the Global Cybersecurity Agenda (GCA) – that strives to be a framework for international cooperation aimed at enhancing confidence and security in the information society. To further develop the GCA a High-Level Experts Group (HLEG) was set up, which finalized the ITU GCA HLEG Report. The GCA collaborates with the International Multilateral Partnership against Cyber Threats (IMPACT) based in Malaysia, which is the largest international public-private cyber security alliance, focusing, inter alia, on early warning systems and developing a global secure electronic collaboration platform for incident response and threat mitigation.

The WSIS also launched the Internet Governance Forum (IGF) in 2005 as a global open multi-stakeholder forum to discuss policy issues related to internet governance. The IGF mandate is to 'discuss public policy issues related to key elements of internet governance in order to foster the sustainability, robustness, security, stability and development of the internet, but also to facilitate discourse and exchange of information and best practices between stakeholders, and to provide advice to those stakeholders. The IGF has no oversight function and is constituted to be a neutral, non-duplicative and non-binding process. The IGF's mandate was extended in December 2015 by the UN General Assembly to 2025. The Secretariat is located in the United Nations Office in Geneva.

## Cyber Warfare and Defence

Cyber warfare involves the actions by a nation-state or an entity to penetrate another nation's computers or information networks for the purpose of causing damage or disruption. It involves warfare conducted in cyberspace through cyber means and techniques rather than conventional means[12]. Cyberspace can be defined as the totality of digital information and communications infrastructures, including the internet, telecommunications networks, computer systems and microprocessors, and the information contained therein. The number of individuals actively using the Internet has grown rapidly to more than 1.7 billion in late 2010. Today, states, non-state communities, business, academia and individuals have become interconnected and interdependent to an unprecedented extent. At the same time, military reliance on computer systems and networks has increased greatly, opening a "fifth" domain of war-fighting next to the traditionally recognized domains of land, sea, air and outer space.

Cyber-espionage like traditional espionage is generally assumed to be ongoing between major powers and is not regarded as an act of war though it can prepare the ground for it. Despite this assumption, some cyber espionage incidents can cause serious tensions between nations, and are often described as hostile actions.

Sabotage through cyber means of computers and satellites that are critical parts of a system or infrastructure can be highly disruptive and damaging. Compromise of military systems that are responsible for orders and communications could lead to their interception or malicious alteration. Power, water, fuel, communications, and transportation infrastructure all may be vulnerable to disruption. In 2010 a malicious software program called Stuxnet infiltrated industry computers around the world, the first case of an attack on critical industrial infrastructure base of modern economies. Denial-of-service attack (DoS attack) or distributed denial-of-service attack (DDoS attack) can make a machine or network resource unavailable such as banks and financial services

The electric power grid is vulnerable to cyberwarfare. In 2009, reports surfaced that China and Russia had infiltrated the U.S. electrical grid and left behind software programs that could be used to disrupt the system. Massive power disruptions caused by a cyber attack could disrupt the economy, distract from a simultaneous military attack, or create a national trauma.

Cyber conflict today usually involves crime or espionage which are not acts of warfare as defined in international law defines it. The lack of international norms for responsible behavior in cyber space reinforces this perception. One way is to define an action in cyberspace that produced the equivalent effect as an armed attack as an act of war. One issue is whether a cyber action must produce physical damage to be regarded as an act of force, or whether other, intangible damage inflicted outside of armed conflict can also be considered a use of force and an act of war.

Computer network warfare is evolving rapidly outpacing technical capabilities to conduct operations and the governing laws and policies. The U.S. military has created a Cyber Command to find and, when necessary, neutralize cyberattacks and to defend military computer networks. The targets include command-and-control systems at military headquarters, air defense networks and weapons systems that require computers to operate.

The distributed nature of internet based attacks means that it is difficult to determine motivation and identify the attacking party, making it unclear when a specific act should be considered an act of war. Examples of cyberwarfare can be found worldwide. In 2008, Russia began a cyber attack on the Georgian government website .In 2008 Chinese hackers attacked CNN as it reported on Chinese repression on Tibet. Responding to these developments, many countries are developing capacity in cyberwarfare and cyberdefence. China has been reported to have greatly expanded its capabilities, challenging the US to the extent that some analysts have proposed an agreement on mutually assured restraint with respect to cyberspace. In the recent US Presidential elections of 2016, Russia has been accused of penetrating the information systems of the Democratic Party and leaking damaging information as a means of influencing the outcome of the elections.

International law is still to clarify what is and is not acceptable in cyberwarfare. NATO has developed the Tallinn Manual, in 2013, as a non-binding study on how international law could apply to cyber conflicts and cyber warfare. The Shanghai Cooperation Organization (which has China and Russia as members) in 2011 proposed to the UN a document called "International code of conduct for information security". In contrast, the United States' approach focuses on physical and economic damage and injury, putting political concerns under freedom of speech. This difference of opinion has led to reluctance in the West to pursue global cyber arms control agreements. More limited efforts may be undertaken, for example talks with Russia over a proposal to limit military attacks in cyberspace including a direct secure voice communications line between the cybersecurity coordinators on both sides. An initiative called the International Convention on Prohibition of Cyberwar in Internet, seeks to make the Internet free from warfare tactics and be treated as the common heritage of mankind.

## Social media and information warfare

The internet has enabled the rapid growth of social media such as facebook, twitter, and numerous communication platforms including on mobile phones. Such media platforms can connect with several million persons, making it possible to communicate instantly across national boundaries. This development has transformed society, politics, marketing and development in a way that is still unfolding. At the same time, it has become easier to spread false and misleading information especially in tense and conflict situations. In August 2012, Indian authorities asked Facebook, Google and Twitter, to remove or block hundreds of pages after political unrest erupted in various parts of the country[13], as Indian northeasterners began fleeing Bangalore, after text messages said to threaten northeasterners were sent around.

Social media is said to have played a big role in the upheavals of the Arab spring in several countries in 2011[14] that led to the downfall of governments[15]. A covert twitter project targeting Cuba called zunzuneo[16] was exposed when USAID funding stopped. Social media destabilization activity has been alleged in Ukraine[17] and Venezuela[18]. This could be regarded examples of information or psychological warfare targeting a country. The Islamic State, Al Qaeda, and similar groups have exploited the potential of social media[19] to influence and mobilize support[20]. This has led some countries to consider regulating social media, to prevent damage, though this path has many technical and political issues. Clearly social media is going to play an increasing role in the future and various groups become proficient in using it and new platforms come into existence.

## Digital manufacturing

The ICT revolution has the potential to radically transform manufacturing. Digital manufacturing is the use of an integrated, computer-based system comprised of simulation, three-dimensional (3D) visualization, analytics and various collaboration tools to create product and manufacturing process definitions simultaneously. The growth in data and new computing capabilities, along with advances in artificial intelligence, automation and robotics, additive technology, and human-machine interaction, are changing the nature of manufacturing itself[21]. Digital-manufacturing technologies will transform every link in the manufacturing value chain, from research and development, supply chain, and factory operations to marketing, sales, and service. Digital connectivity among designers, managers, workers, consumers, and physical industrial assets will unlock enormous value and change the manufacturing landscape forever.

Enterprises and governments are reacting to this change. Manufacturers are starting to use data analytics to optimize factory operations, boosting equipment utilization and product quality while reducing energy consumption. Managers have a clearer view of raw materials and manufactured parts flowing through a manufacturing network, which can help them to optimize operations to cut costs and improve efficiency. Smart, connected products are sending customer experience data to product managers to help them anticipate demand and maintenance needs and design better products. The digital manufacturing revolution is under way. New technologies such as nanotechnology will also have a significant impact. Leaders in digital manufacturing, including some smaller players, are already gaining significant competitive advantage.

The National Network for Manufacturing Innovation in the US[22] is organizing six major research institutes of which one is focused specifically on digital manufacturing. Similar efforts are underway across the globe, including Germany's Industry 4.0 effort and China's made in China 2025. A global convening organization, the Industrial Internet Consortium, was founded just 18 months ago and already has 175 members.

Digital manufacturing technology will be disruptive[23]. Fewer employees will be needed due to automation and more employees will be needed with skills in design, engineering, IT, logistics, marketing and other professions. The manufacturing jobs of the future will require more skills. The revolution will affect not only how things are made, but where. For decades, low-wage countries have used manufacturing as an engine of prosperity, luring production from abroad with cheap labour costs. Digital manufacturing is halting the flow of production to countries where labour costs are low and in some cases is even bringing production back to the US. Labour costs are growing less and less important: and production is increasingly moving back to rich countries because companies now want to be closer to their customers so that they can respond more quickly to changes in demand. And some products are so sophisticated that it helps to have the people who design them and the people who make them in the same place. It is estimated that in areas such as transport, computers, fabricated metals and machinery, 10-30% of the goods that America now imports from China could be made at home by 2020, boosting American output by $20 billion-55 billion a year.

## Telemedicine

Telemedicine is the use of medical information exchanged from one site to another via electronic communications to improve a patient's clinical health status[24]. It helps eliminate distance barriers and can improve access to medical services that would often not be consistently available in distant rural communities. Telemedicine uses a growing variety of applications and services using two-way video, email, smart phones, wireless tools and other forms of telecommunications technology. The use of telemedicine has spread rapidly and is now becoming integrated into the ongoing operations of hospitals, specialty departments, home health agencies, private physician offices as well as consumer's homes and workplaces.

Products and services related to telemedicine are often part of a larger investment by healthcare institutions in either information technology or the delivery of clinical care. Patient consultations via video conferencing, transmission of still images, e-health including patient portals, remote monitoring of vital signs, continuing medical education, consumer-focused wireless applications and nursing call centers, among other applications, are all considered part of telemedicine and telehealth.

The term telehealth is sometimes used to refer to a broader definition of remote healthcare that does not always involve clinical services. The term health information technology (HIT) refers to electronic medical records and related information systems while telemedicine refers to the actual delivery of remote clinical services using technology.

Telemedicine can be beneficial to patients in isolated communities and remote regions, who can receive care from doctors or specialists far away without the patient having to travel to visit them. Recent developments in mobile collaboration technology can allow healthcare professionals in multiple locations to share information and discuss patient issues as if they were in the same place. Remote patient monitoring through mobile technology can reduce the need for outpatient visits and enable remote prescription verification and drug administration oversight, potentially significantly reducing the overall cost of medical care. Telemedicine can also facilitate medical education by allowing workers to observe experts in their fields and share best practices more easily.

The drawbacks of telemedicine include the cost of telecommunication and data management equipment and of technical training for medical personnel who will employ it. Virtual medical treatment also entails potentially decreased human interaction between medical professionals and patients, an increased risk of error when medical services are delivered in the absence of a registered professional, and an increased risk that protected health information may be compromised through electronic storage and transmission. Poor quality of transmitted records, such as images or patient progress reports can compromise the quality of patient care. Other obstacles to the implementation of telemedicine include unclear legal regulation for telemedical practices and difficulty claiming reimbursement from insurers or government programs. Another disadvantage of telemedicine is the inability to start treatment immediately.

In India the potential of telemedicine for improving health care delivery has been well recognized. A number of initiatives have been undertaken by government as well as private health institutions[25]. A National Task Force on Telemedicine was set up in 2005 to address various issues in telemedicine. The Indian Space Research Organization (ISRO) has set up a satellite based Telemedicine Network consisting of 245 Hospitals – 205 Remote/Rural/District Hospital/Health Centers connected to 40 Super Specialty Hospital located in the major cities. Ministry of Health has started the Onconet programme involving a network connecting 25 Regional Cancer Centres and 100 peripheral centres to provide comprehensive cancer treatment facilities and carry out cancer prevention and research activities. A tele-ophthalmology project has been started to provide eye care specialty services in 3 states of India. Recently, a pan-India health initiative called Sehat[26] was launched in 2015 as part of the Digital India initiative. This effort which will be run in collaboration with Apollo Hospitals, aims to connect 60,000 common service centres across the country and provide healthcare access to citizens irrespective of their geographical location.

Mauritius's Medical Centre has been connected to seven universities and 12 highly specialized hospitals in India which will provide e-health services[27]. India has also set up a SAARC telemedicine network[28], covering the South Asian region, and a project for providing e-health services from 12 super specialty hospitals for 53 African countries through the Pan-African e network project[29].

Given the advantages of e-health services, this area is likely to witness further rapid development, especially if international agreements can be reached on regulatory and licensing issues.

## Bioinformatics

Bioinformatics is the application of computer technology to the management of biological information. Computers are used to gather, store, analyze and integrate biological and genetic information which can then be applied to gene-based drug discovery and development. The genomic information resulting from the Human Genome Project has led to rapid development. Universities, government institutions and pharmaceutical firms have formed bioinformatics groups, consisting of computational biologists and bioinformatics computer scientists. Such groups will be key to unraveling the mass of information generated by large scale sequencing efforts underway in laboratories around the world.

As an interdisciplinary field of science, bioinformatics combines computer science, statistics, mathematics, and engineering to analyze and interpret biological data. Bioinformatics techniques allow extraction of useful results from large amounts of raw data. It is applied in sequencing and annotating genomes and their observed mutations, organizing biological data, analysis of gene and protein expression and regulation, comparison of genetic and genomic data, analysis of biological pathways and networks and simulation and modeling of DNA, RNA, and protein structures as well as molecular interactions.

A number of Indian institutions government and private have developed capability in bioinformatics research and education. The Departments of Information Technology and of Biotechnology have adopted promotional and support programmes in this field. These include the Biotechnology Information System Network (BTISNet) since 1986, establishment of nine bioinformatics Centres in India, etc. The bioinformatics sector in India[30] is expected to show vigorous growth due to genomics, translational bioinformatics, and personalized medicine, besides providing outsourcing to foreign associates, and leveraging of India's large IT skilled human resources. Other trends in this sector that offer unlimited opportunities include structural biology services, computer-assisted molecule discovery, and bio- and chemo-informatics. Bioinformatics is a rapidly expanding sector globally as well as in India.

## Outlook

ICT is likely to continue to develop rapidly in the future, with smaller and more powerful processing devices becoming possible through developments in nanotechnology. Artificial intelligence will also develop and open up new possibilities. The increasing spread of computing systems including embedded systems in "smart" household, and workplace devices will continue. This will increase the vulnerability to cyber crime and cyber warfare, requiring new and more effective responses. These challenges will require diplomacy to handle cross border ramifications.

# Endnotes

1 History of ICT, ITU website, http://www.itu.int/en/ictdiscovery/Pages/timeline.aspx accessed 29-1-2016

2 History Of Information Technology In India, N. Sheike, 22 Apr 2012, http://www.indiastudychannel.com/resources/151102-History-Information-Technology-India.aspx accessed 29-1-2016

3 Modes of Supply for International Services, US International Economic Accounts Chapter 14, https://www.bea.gov/international/pdf/concepts-methods/14%20Chapter%20ITA-Methods.pdf accessed 30-1-2016

4 The General Agreement on Trade in Services (GATS): objectives, coverage and disciplines, WTO, https://www.wto.org/english/tratop_e/serv_e/gatsqa_e.htm accessed 30-1-2016

5 INDIA-State of Play in Services under the GATS at WTO, Ministry of ommerce, Government of India, 2014, http://commerce.nic.in/trade/INDIA_State_of_Play_in_Services%20_under_the_GATS_at_WTO.pdf accessed 30-1-2016

6 Trade in Services Agreement (TiSA), European Commission, http://ec.europa.eu/trade/policy/in-focus/tisa accessed 30-1-2016

7 Free Trade Agreements-Frequently Asked Questions (FAQs), Ministry of Commerce, India, http://commerce.nic.in/trade/FAQ_on_FTA_9April2014.pdf?id=9&trade=i accessed 31-1-2016

8 What is Cyber Security? http://www.itgovernance.co.uk/what-is-cybersecurity.aspx

9 Cybersecurity overview, US Department of Homeland Security, http://www.dhs.gov/cybersecurity-overview accessed 31-1-2016

10 UNODA Fact Sheet July 2015 https://unoda-web.s3.amazonaws.com/wp-content/uploads/2015/07/Information-Security-Fact-Sheet-July2015.pdf accessed 31-1-2016

11  An Updated Draft of the Code of Conduct, Incyder news, February 2015, https://ccdcoe.org/updated-draft-code-conduct-distributed-united-nations-whats-new.html accessed 1-2-2016

12  Cyberwarfare and International Law 2011, Nils Melzer, United Nations Institute for Disarmament Research (UNIDIR), http://unidir.org/files/publications/pdfs/cyberwarfare-and-international-law-382.pdf accessed 1-2-2016

13  India Blocks Facebook, Twitter, Mass Texts in Response to Unrest, Mediashift.org, 28 August 2012, http://mediashift.org/2012/08/india-blocks-facebook-twitter-mass-texts-in-response-to-unrest241 accessed 1-2-2016

14  New study quantifies use of social media in Arab Spring, UW Today, 12 September 2011, http://www.washington.edu/news/2011/09/12/new-study-quantifies-use-of-social-media-in-arab-spring accessed 1-2-2016

15  From Arab Spring to Autumn Rage: The Dark Power of Social Media, Huffington Post 14 September 2012, http://www.huffingtonpost.com/entry/social-media-middle-east-protests-_b_1881827.html accessed 1-2-2016

16  Senate committee probes 'Cuban Twitter' USAid ZunZuneo programme, The Guardian, 10 April 2014, http://www.theguardian.com/world/2014/apr/10/senate-committee-cuban-twitter-usaid-zunzuneo accessed 1-2-2016

17  The Twitter War: Social Media's Role in Ukraine Unrest, National Geographic, 11 May 2014, http://news.nationalgeographic.com/news/2014/05/140510-ukraine-odessa-russia-kiev-twitter-world accessed 1-2-2016

18  Social Media Coup? The Vile Virality of Venezuela's Opposition, Venezuelaanalysis.com 17 February 2015 http://venezuelanalysis.com/analysis/11213 accessed 1-2-2016

19  ISIS, Al Qaeda Exploit Social Media To Spread 'Distorted Ideology,' Internet Companies Must Respond: UN, International Business Times,

25 June 2015, http://www.ibtimes.com/isis-al-qaeda-exploit-social-media-spread-distorted-ideology-internet-companies-must-1982937 accessed 1-2-2016

20 FBI chief: Isil's social media strategy makes it bigger threat to US than al-Qaeda, The Telegraph, 23 June 2015, http://www.telegraph.co.uk/news/worldnews/islamic-state/11758558/FBI-chief-Isils-social-media-strategy-makes-it-bigger-threat-to-US-than-al-Qaeda.html accessed 1-2-2016

21 Digital manufacturing: The revolution will be virtualized, Mc Kinsey & Company, August 2015, http://www.mckinsey.com/insights/operations/digital_manufacturing_the_revolution_will_be_virtualized accessed 1-2-2016

22 Enabling Digital Manufacturing: A Strategy to Develop a National Innovation Network , National Center for Manufacturing Sciences (NCMS), 30 September 2010, http://www.ncms.org/wp-content/uploads/2012/07/Revitalizing-Manufacturing-DM-Whitepaper-093010.pdf accessed 1-2-2016

23 The third industrial revolution, The Economist, 21 August 2012, http://www.economist.com/node/21553017 accessed 1-2-2016

24 What is telemedicine? American Telemedicine Association, http://www.americantelemed.org/about-telemedicine/what-is-telemedicine accessed 2-2-2016

25 E-health initiatives in India, telemedindia http://telemedindia.org/india/e-Health%20Initiatives%20in%20India.pdf accessed 2-2-2016

26 Govt launches Sehat telemedicine initiative, Live Mint, 25 August 2015 http://www.livemint.com/Politics/df1ef1TTLuegLCj7g9qWRK/Govt-launches-Sehat-telemedicine-initiative.html accessed 2-2--2016

27 Mauritius Medical Centre to be linked to 12 Indian hospitals, 16 September 2008 by eHEALTH, http://ehealth.eletsonline.com/2008/09/mauritius-medical-centre-to-be-linked-to-12-indian-hospitals accessed 2-2-2016

28 SAARC Telemedicine Project in India, http://www.saarctf.org/ Countries/India.aspx accessed 2-2-2016

29 Pan-African e-Network Project, http://www.panafricanenetwork.com accessed 2-2-2016

30 Indian Bioinformatics: Growth Opportunities and Challenges Dr. Pankaj Madhani, ResearchGate, 2011, https://www.researchgate. net/publication/228294796_Indian_Bioinformatics_Growth_ Opportunities_and_Challenges accessed 2-2-2016

# Chapter 8

# Aerospace Technology – From the Earth to the Planets and beyond

*"Centuries hence, when current social and political problems may seem as remote as the problems of the Thirty Years' War are to us, our age may be remembered chiefly for one fact: It was the time when the inhabitants of the earth first made contact with the vast cosmos in which their small planet is embedded."*

— Carl Sagan, Scientific American magazine, March 1975

## Introduction

Aerospace technology is the technology of flight in the atmosphere of Earth and surrounding space. Airspace is the physical air space directly above a location on the ground and is considered as ending 100 km above the ground where the air pressure becomes too low. Beyond this distance is what is termed space. Aerospace technology involves research, design, and manufacturing, operating, or maintaining aircraft and/or spacecraft. Aerospace activity is very diverse, with a multitude of commercial, industrial and military applications. The design of a flight vehicle is highly demanding in terms of knowledge of many engineering disciplines such as aerodynamics, propulsion systems, structural design, materials, avionics, and stability and control systems.

Mastery over airspace became a strategic necessity for all countries. This required the use of advanced aircraft, detection and defensive systems. Most countries include air force components in their defence forces and military strategy. Countries that cannot produce military aircraft and systems need to purchase these systems from those that do. Nuclear weapons states have over time developed longer and longer range missiles to deliver nuclear warheads far away. As missile launch vehicles developed, it became possible also to launch satellites of increasing weight and orbital distance. Satellites in space added a new dimension to military strategy. The aerospace sector thus assumed importance in both civilian and strategic fields. Mastery of aerospace industry required a synthesis of many disciplines – such as materials and structures, aeronautics, fluid dynamics, rocket and jet propulsion, computers and avionics, etc. and extreme reliability and precision. Only a few countries could reach such a level of technological capability to produce complete aerospace systems, and this gave them an advantage as suppliers to other countries. The emergence of the era of aerospace and air transport also brought with it attempts by governments to regulate airspace and outer space, through new treaties and legal instruments. Disputes over airspace boundaries also arose between nations states, especially over the seas.

## Historical Development of Aerospace Technology

The roots of aerospace technology can be traced to sketches of flight vehicles drawn by Leonardo da Vinci. Manned flight was first achieved in 1783, in a hot-air balloon designed by the French brothers Joseph-Michel and Jacques-Étienne Montgolfier. The power-driven balloon was invented and operated by Henri Giffard, a Frenchman[1], in 1852. In 1799 Sir George Cayley, an English baron, drew an airplane incorporating a fixed wing for lift, an empennage (consisting of horizontal and vertical tail surfaces for stability and control), and a separate propulsion system[2]. The newly invented gasoline engine and propeller were added by the Wright brothers to existing glider design. On December 17, 1903, their machine rose from the track and covered 120 feet in about 12 seconds. Following the first sustained flight of a heavier-than-air vehicle in 1903, the Wright brothers refined their design, but struggled to sell the idea to a skeptical US Army. They eventually succeeded in 1908 following President Theodore Roosevelt's intervention.

The major impetus to aircraft development and aerospace technology occurred during World War I and later in World War II. The role of air power in military operations provided a strong incentive for development, and aircraft were designed and constructed for specific military missions, including fighter attack, bombing, and reconnaissance. During World War II, the rival powers made major efforts in aircraft development without regard to cost. The Germans first deployed unmanned cruise and ballistic missiles using jet engines in 1944[3]. The modern jet engine was invented by Frank Whittle in 1930 and by May 1941, the first flight using the jet engine was successfully made. The era of jet aircraft was launched. By 1956, the supersonic speed of bombers and fighters made them practically indistinguishable, each with about the same weight and the same operational ceiling.

During the 1920s, aircraft assumed their modern shape. Monoplanes replaced biplanes, and radial air-cooled engines turned variable pitch propellers, and enclosed fuselages and cowlings gave aircraft their aerodynamic shape. By the mid-1930s, metal replaced wood as the material of choice in aircraft construction. Users such as the military formed air units specifically to exploit this new technology. Air transport companies began flying passengers in the 1920s, though all those airlines were kept afloat by government airmail contracts. European nations developed airmail routes around their colonies. The United States was the only country with a large indigenous airmail system, and it drove the structure of the industry during the 1920s.

The United States being a large country, developed air travel, and by the 1930s airlines competed for passengers by forging alliances with aircraft manufacturers. The Douglas DC-3, introduced in 1935, gave airlines efficient means of carrying people rather than mail. Many advances in aircraft design during the 1930s addressed the comfort, efficiency and safety of air travel — cabin pressurization, retractable landing gear, better instrumentation and better navigational devices around airports.

Britain and Germany produced the best large bombers at the start of the 1930s. European states had stepped up production of military aircraft, training pilots to fly them, and building airfields to host them. Once the war began, factories were bombed and supply lines cut off. German and British aircraft firms instead invested in research and engineering to create better aircraft. Europeans developed the strategic missile, the jet engine, better radar, all-weather navigation aids, and more nimble fighters. The German Messerschmitt 262 fighter aircraft, which combined a strong turbine engine with the innovation of swept wings, approached the speed of sound. The Europeans also innovated in tactics and logistics to use fewer aircraft more effectively.

American firms were advanced in mass production. In the six-year period 1940 through 1945, American firms built huge numbers of military aircraft and the aviation industry became America's largest producer and employer. A vast array of firms including subcontractors, became involved in manufacturing, and for the next half-century Americans set the agenda for the aircraft industry around the world.

In civilian aviation, De Havilland was the first to develop a jet passenger liner. By May 1952, one was available with 48 passengers going 800 kilometres per hour and flying at 42,000 feet. Today jet airliners are commonly used, carrying a large volume of passenger and air cargo all over the globe.

Jet and rocket engines had an almost parallel development. An American Professor of Physics, Robert H. Goddard, successfully tested rockets and was granted two patents on rocketry in 1914. Rockets were developed by Germany during the World War II to deliver bombs to cities in Britain, but were of little military value. At the end of the War, many of the scientists and engineers involved moved to the US where they continued work on rockets and missiles. As nuclear weapons became smaller and lighter, missiles with increasing accuracy and range became the basis of nuclear deterrence between the US and the USSR.

## Aerospace Industry and the Cold War

The Berlin airlift of 1947 marked the start of the Cold War between the United States and the Soviet Union, in which perceptions of aerial might played a key role. Aircraft designs integrated the technological advances of World War II. Several countries such as the US, USSR, France, and UK competed to develop faster, higher flying and more maneuverable military aircraft equipped with more and more advanced electronic systems and weapons. Air defence systems using guns and missiles controlled by radar were also developed. The most advanced aircraft producing countries, the US and USSR became suppliers to the air forces of their allies, and military aircraft and air defence systems sales to countries became a way of expanding political influence.

A new type of airframe known as a helicopter gained prominence for both civilian and military applications. Electronics firms became prime contractors for new guided missiles, while airframe manufacturers subcontracted to them. Turbojet engines were a disruptive new technology replacing piston engines. Firms such as McDonnell Aircraft and Lockheed developed airframes for the greater speeds and altitudes possible with jet engines. Intercontinental ballistic missile (ICBM) programs, started in 1954, further stimulated the industry. The complexity of the designs, the reliability required of each part, and the speed for which the missiles had to be designed and built, were major challenges to be overcome.

Also revolutionary were the spacecraft and the rockets that lifted missiles into space. The term "aerospace" reflected the situation following the Soviet launch of Sputnik in October 1957, which galvanized the US race into space. The US National Advisory Committee for Aeronautics was the basis of the new National Aeronautics and Space Administration, NASA, which then promoted efforts of academic aeronautics toward hypersonics and space travel. In 1961, NASA was tasked to send an American to the Moon and return him safely to Earth before the end of the decade. NASA built enormous space ports in Florida and Texas, enhanced its arsenal of research laboratories, bolstered its own network of hardware contractors, opened up new areas of material science, and pioneered new methods of reliability testing. Following the success of Apollo mission, NASA started a programme to create the space shuttle for regular access to space.

## Aviation Law and governance

Prior to World War I, several nations signed bilateral agreements regarding the legal status of international flights, and during the war, several nations took the step of prohibiting flights over their territory. Several competing multilateral treaty regimes were established in the wake of the war. The lack of uniformity in international air law, particularly with regard to the liability of international airlines, led countries to agree on the Warsaw Convention of 1929. The International Air Transport Association (IATA) was founded in 1919 to foster cooperation between airlines in various commercial and legal areas. Aviation law developed as a branch of law that encompasses most facets of air travel, as well as the operation and regulation of business issues relating to air travel, which requires a comprehensive knowledge of aviation regulations, specific laws regarding flight, and an in depth understanding of aviation. Aviation law has both international and national aspects and due to the rapid expansion of air services, countries have recognized the need for cooperation.

The Chicago Convention on International Civil Aviation was signed in 1944, during World War II, and provided for the establishment of the International Civil Aviation Organization as the central agency of the United Nations system devoted to overseeing civil aviation. The Convention also provided various general principles governing international air services. Responding to the threats of hijacking of aircraft, the Tokyo Convention of 1963 set up new international standards for the treatment of criminal offenses on or involving aircraft. The Montreal Convention of 1999 updated the carrier liability provisions of the Warsaw Convention, while the Cape Town Treaty of 2001 created an international regime for the registration of security interests in aircraft and certain other large movable assets.

The International Civil Aviation Organization (ICAO) is a UN specialized agency[4], established by States in 1944 to administer the Convention on International Civil Aviation (Chicago Convention). ICAO works with the Convention's 191 Member States and aviation industry groups to reach consensus on international civil aviation Standards and Recommended Practices (SARPs) and policies in support of a safe, efficient, secure, economically sustainable and environmentally responsible civil aviation sector. These SARPs and policies are used by ICAO Member States to ensure that their local civil aviation operations and regulations conform to global norms, which in turn permits more than 100,000 daily flights in aviation's global network to operate safely and reliably in every region of the world. In addition to its core work of evolving consensus-driven international SARPs and policies, ICAO also coordinates assistance and capacity building for States in support of numerous aviation development objectives; produces global plans to coordinate multilateral strategic progress for safety and air navigation; monitors and reports on numerous air transport sector performance metrics; and audits States' civil aviation oversight capabilities in the areas of safety and security.

The International Air Transport Association (IATA) is the trade association for the world's airlines, representing some 260 airlines or 83% of total air traffic. It supports many areas of aviation activity and helps formulate industry policy on critical aviation issues.

## Rocketry and Space exploration[5]

The Cold War and pursuit of strategic interests stimulated a race between the US and the USSR in the field of upper atmosphere and space research. The USSR took the lead, with the launch of the first satellite Sputnik 1 in 1957, which jolted the USA into the space race. Important landmarks were the first human to orbit the Earth (Yuri Gargarin, USSR, 1961), the first human to land on the moon (Apollo 11, USA, 1969), followed by interplanetary landings on Mars and Venus. The era of space exploration had taken off. More countries joined the race, including the European Union, China, Japan, and India. Over the decades, the focus shifted to renewable space exploration vehicles such as the space shuttle programme (USA, 1981-2011) and from competition to cooperation in projects such as the International Space Station ( 1988 to date), being the largest artificial body in earth orbit continuously inhabited by humans for over 15 years. In recent years, space exploration has steadily pushed outwards to the outer planets of the solar system, and beyond into deep space, revealing secrets of the origin of the solar system and the evolution of planets and their satellite systems.

## Satellite Applications- Civilian and Military

The spin offs from space exploration were apparent in both civil and military domains. In the civilian field, space based platforms provided the basis of reliable, high capacity communication systems spanning the globe, direct television broadcasting of high quality and reliability, and the ability to gather data on the earth's surface, atmosphere and oceans on an unprecedented scale and detail. The applications of these were wide ranging, from telecommunications, the internet, meteorology, human habitation, agriculture, etc.

In the military domain, space based platforms offered similar capacities such as global communication systems, information collection and reconnaissance of adversary activities, etc. Space based platforms became closely integrated into cyber networks used by the military, leading to the possibility of space based platforms becoming targets in warfare. Offensive anti satellite weapons (ASATs) of various types (kinetic and non kinetic) are now being tested and may well become part of future arsenals. Complicating the scenario is the fact that many space based platforms are used both for critical civilian applications as well as for military applications.

There were other spin offs from space exploration efforts. These included electronic cameras using high resolution CCD devices, new materials, etc. Some of the product s generated by the spin offs are now widely used in everyday life.

## Arms Control and Space Law issues

The possibility of war in outer space had been foreseen since the early days of the space race. Early on international consensus was reached on a treaty banning the testing, deployment and use of nuclear weapons in outer space, the so called "Outer Space Treaty". By then some nuclear tests in outer space had been carried out to examine the possibility of using the resulting electromagnetic pulse (EMP) to knock out electrical systems. The results did not indicate much military promise. The Outer Space Treaty does not prevent the launch and transit of nuclear weapons in outer space, as this would end the use of ballistic missiles by the nuclear weapons states.

A recent development is the effort by leading space nations such as the US, Russia and China to develop anti satellite weapons and systems. The ASAT race has begun, promising to destabilize the space domain. ASATs can physically destroy target satellites by impact or by explosives, or be designed to rendezvous with the target and disable its systems. Testing of ASATs has further added to the growing amount of "space junk" that jeopardizes space vehicles. China's ASAT test of 2007 achieved the dubious distinction of the world's biggest "space junk" incident, creating over 1 million pieces, with over 35000 over 1 cm in size. An even more daunting prospect is a relatively crude ASAT weapon that deliberately produces a large cloud of space junk to attack a satellite. Having achieved a degree of advancement in this field, the three leading ASAT players are now seeking to close the door to others by proposing a Non-ASAT proliferation treaty, similar to the NPT, while preserving and enhancing their own ASAT systems. Such is the hypocrisy in some arms control efforts which are more in the nature of arms legitimizing regimes for the dominant players.

There is a growing body of space law which is the international law applicable in outer space (beyond 100 km from the earth's surface). The most important of these is under the auspices of the UN's Committee on the Peaceful uses of Outer Space[6] consisting of five Treaties and five declarations of principles. The Outer Space Treaty, 1967, represents the basic legal framework of international space law. It bans the placing of weapons of mass destruction in orbit of Earth, installing them on the Moon or any other celestial body, or otherwise stationing them in outer space. It exclusively limits the use of the Moon and other celestial bodies to peaceful purposes and expressly prohibits their use for military purposes. However, the Treaty does not prohibit the placement of conventional weapons in orbit, a huge strategic loophole, which enables ASATs to be developed. The treaty also states that the exploration of outer space shall be done to benefit all countries and shall be free for exploration and use by all the States.

The treaty explicitly forbids any government from claiming a celestial resource such as the Moon or a planet, claiming that they are the common heritage of mankind, and states that outer space is not subject to national appropriation by claim of sovereignty, by means of use or occupation, or by any other means. However, the State that launches a space object retains jurisdiction and control over that object and is also liable for damages caused by their space object.

Other international agreements are the Rescue Agreement (1968), the Liability Convention (1972), the Registration Convention (1975), and the Moon Agreement (1979). The five declarations of principles adopted by the UN are the Declaration of Legal Principles (1963), the Broadcasting Principles (1982), the Remote Sensing Principles (1986), the Nuclear Power Sources principles (1992), and the Benefits Declaration (1996). These are non-binding in nature, and reflect the fact that it is becoming increasingly complex to negotiate agreements on issues that are rapidly evolving. For example, the Broadcasting Principles require that for direct broadcasting services, the agreement of the broadcast transmitting and receiving state is necessary, unlike radio broadcasting where there is "freedom of the air waves"" and only technical regulation by the ITU.

To make some order out of the rapidly growing number of objects launched into outer space, the Convention on Registration of Objects Launched into Space, 1976 requires member states to maintain national registers of such objects, which are made available through the UN. This system covers about 92% of the objects launched into space. However, there is a huge growth in "space junk" or debris[7] from various space launches, dead satellites, boosters, and fragments from collisions, amounting to some 170 million objects including some 29000 large objects ( over 10 cm size). These pose a serious and growing hazard to space vehicles, as more and more space activities take place.

## Missile Technology and Control Regimes

The development of space exploration resulted in development of large space launch vehicles capable of launching objects of increasing weight to greater distances into space. The same technology could also launch nuclear weapons across the earth in the form of ballistic missiles of increasing range. In fact ballistic missile launch capability and space launch capability are two sides of the same coin, a typical dual use technology. The countries in the forefront of the space race, especially the US and its allies wished to keep access to this new technology limited to their allies and prevent adversaries from acquiring this technology. At the same time, export of some technology would be necessary to allow for commercial development. Therefore a limit was set of 500 kg payload and 300 km range for missile technology, assuming that this would prevent spread of missile technology for delivering nuclear weapons. There were serious flaws in this reasoning. Nuclear weapons could be developed with less weight than 500 kg, and range could be increased beyond 300 km if the payload was lighter. Later the objective of the arrangement was to prevent delivery systems for "weapons of mass destruction", i.e. chemical, biological, or radiological weapons. However such weapons could be developed with a weight much lower than 500 kg. It would have been more logical to focus on restricting the end use of the launch vehicles, even though verification and compliance might be problematic.

The Missile Technology Control Regime (MTCR) is a multilateral export control regime established in 1986 by the G7 countries. It is an informal and voluntary partnership among 35 countries (including India as of 2016) to prevent the proliferation of missile and unmanned aerial vehicle technology capable of carrying above 500 kg payload for more than 300 km. The scope was expanded to include nonproliferation of unmanned aerial vehicles (UAVs) for all weapons of mass destruction. Prohibited materials are divided into two Categories, which are outlined in the MTCR Equipment, Software, and Technology Annex. Israel, Romania and Slovakia have agreed to voluntarily follow MTCR export rules even though not yet members. India has joined the MTCR in 2016. China pledged in 1991 that it would adhere to the MTCR, but its membership application is pending since 2004 over concerns over its export control regime. However, missile technology proliferation continues with China, North Korea, and Iran providing technology and missiles to Pakistan, Saudi Arabia and Syria, respectively.

## India's aerospace development and issues

India's first aerospace enterprise, Hindustan Aircraft Co. was set up as a private entity in 1940.[8] During World War II, the British Government of India nationalized it and greatly expanded its operations as a strategic necessity to provide engineering support to allied air forces. After independence, the Government of India took it over and in 1964 after consolidation of aircraft manufacturing units, it became Hindustan Aeronautics Ltd. (HAL). HAL became the dominant aerospace player in India, expanding its operations to include military aircraft, helicopters, and engines. During the 1980s, HAL developed indigenous aircraft such as the HAL Tejas and HAL Dhruv. HAL also developed an advanced version of the Mikoyan-Gurevich MiG-21, which increased its life-span by more than 20 years. HAL has also obtained several multimillion-dollar contracts from leading international aerospace firms such as Airbus, Boeing and Honeywell to manufacture aircraft spare parts and engines.

However, in recent years, the demands for aerospace products in India both in civil and military sectors has grown tremendously, and far exceeds HAL's capacity. India is therefore much more dependent on import of foreign aerospace products and systems. Recently the Government has liberalized its policies to enable the private sector domestic as well as foreign to play a greater role in aerospace manufacturing. India's capacity for R & D and development of cutting edge aerospace products has fallen behind countries such as China, causing concerns among the strategic community.

India's space programme was launched under the Department of Atomic Energy in 1950. The Indian Space Research Organization (ISRO) has become the dominant entity. It emerged in 1969 out of its predecessor the INCOSPAR, and is managed by the Department of Space of the Government of India. In 1980, the first Indian satellite was placed in orbit by an Indian-made launch vehicle, SLV-3. ISRO subsequently developed two other rockets: the Polar Satellite Launch Vehicle (PSLV) for launching satellites into polar orbits and the Geosynchronous Satellite Launch Vehicle (GSLV) for placing satellites into geostationary orbits. These rockets have launched numerous communications satellites and earth observation satellites. Satellite navigation systems like GAGAN and IRNSS have been deployed. The first GSLVs were built by India with the cryogenic engine purchased from Russia, but technology transfer was cut off under the Missile Technology Control Regime. This resulted in ISRO having to make an intensive effort to develop the technology for the GSLV.

In January 2014, after several failures since 2010, ISRO successfully used an indigenous cryogenic engine in a GSLV-D5 launch of the GSAT-14. In 2008, ISRO sent one lunar orbiter, Chandrayaan-1, and in 2014 placed a Mars orbiter into a Mars orbit, making India the first nation to succeed on its first attempt, and ISRO the fourth space agency in the world to successfully reach Mars orbit. Future plans include development of GSLV Mk III,(for launch of heavier satellites), unified launch vehicles, development of a reusable launch vehicle, human spaceflight, further lunar exploration, interplanetary probes, and a solar spacecraft mission. India is developing a supersonic combustion ram (SCRAM) jet [9] for use in augmenting apace launch vehicle capability. In space applications, India has launched a large number of satellites, including geo-stationary satellites for communications (INSAT), remote sensing satellites (IRS), radar imaging satellites (RIS), and global navigational satellites (IRNSS). Space exploration activities include satellites orbiting the Moon and Mars.

Military applications of space, particularly development of missiles was shelved off from ISRO to the Defence Research and Development Organization (DRDO). DRDO carried out the integrated missile development programme from 1980, which gave birth to a family of short, medium and long range missiles with capability for launch from various platforms. Its Agni IV solid fuelled ICBM has a range of up to 8000 kilometres, and longer range versions are under development. DRDO also develops advanced aircraft development, unmanned aerial vehicles (UAVs), and avionics and collaborates closely with HAL. With a network of 52 laboratories covering aeronautics, armaments, electronics, land combat engineering, life sciences, materials, missiles, and naval systems, etc. DRDO is India's largest and most diverse research organization with around 5,000 scientists. Indian scientists have also reportedly developed a new technology[10] that would reduce atmospheric drag on missiles and space launchers by 40 percent, enabling missile capacity to be increased significantly.

An India-Russia enterprise, Brahmos Aerospace has developed the world's first supersonic cruise missile the Brahmos with a range of 290 km and a top speed of Mach 3. This missile has been deployed on Indian surface warships and air and submarine launched versions are being developed. Several countries such as Malaysia and Vietnam[11] have shown interest in acquiring this missile for their warships and aircraft. The Brahmos had to be developed with a restricted range due to MTCR restrictions. With India joining the MTCR in 2016, this limitation no longer applies, and longer range cruise missile programme could certainly be developed. The Brahmos project could be a precedent and useful model for joint ventures with other countries for defence manufacturing, given India's large requirements.

## International collaboration in Space

As space projects become increasingly complex and expensive, countries have sought to work together on such projects. The European Space Agency (ESA), set up in 1975[12], now has 22 European nations as members, and a budget of over Euro 5 billion. Its purpose is defined as "to provide for, and to promote, for exclusively peaceful purposes, cooperation among European States in space research and technology and their space applications, with a view to their being used for scientific purposes and for operational space applications systems". It works closely with the European Union, and has successfully launched satellites into space. Its Ariane 5 launcher has been operating commercially as a space launch vehicle for satellites of many countries. The ESA cooperates with space agencies of many non-member states. Canada is a cooperating state, and the newer members of the EU have signed various agreements with ESA with the aim of ultimately joining it. The ESA offers a useful example of intergovernmental cooperation in large scale scientific programmes where no one country can afford to carry out large projects.

Over the years collaboration has gradually grown in space activities, replacing competition. The International Space Station (ISS) Program is an example of international cooperation[13] among the space agencies of the United States, Russia, Europe, Japan, and Canada. The ISS has been the most politically complex space exploration program ever undertaken. Launched in 1998 and involving the U.S., Russia, Canada, Japan, and the participating countries of the European Space Agency, the International Space Station is the largest space station ever constructed. It continues to be assembled in orbit. It has been visited by astronauts from 18 countries.

An International Space Exploration Coordination Group (ISECG)[14] was set up in 2006 as a forum to coordinate the efforts of the 14 participating space agencies in space exploration. In 2014, this was widened when the US hosted the first meeting of the International Space Exploration Forum (ISEF), with 30 countries participating[15]. These are important steps in evolving a framework for international cooperation in space exploration. Hopefully such a framework would result in avoiding conflict and making best use of resources, capabilities and exploiting synergies and complementarities in space exploration. Exploration of the solar system and beyond to the nearest stars will require massive efforts that can be best achieved through international cooperation.

## Commercialization of Space

With the growth in space exploration activities, commercial applications have followed suit. The first important sector was the use of satellite generated data on various earth activities, such as for meteorology, mapping , and surveying for resources. Satellite based communications have grown rapidly and now cover direct radio and television broadcasting, as well as for telephone and data communications. Another important segment is space transportation, which includes the launching of satellites, cargo and humans into space. A third segment is the manufacture of satellites and space systems and equipment, which may overlaps in some respects with traditional aeronautical industry. A fourth area is the commercial recovery of space resources, for example mining of the Moon and asteroids. These possibilities have now moved from the realm of science fiction to the plausible future. The future growth of commercialization of outer space will require appropriate regulatory responses.

The issue of ownership of property in space has been an untested area, with strong arguments both for and against. The making of national territorial claims in outer space and on celestial bodies has been specifically proscribed by the Outer Space Treaty. It has been suggested that the lack of private property rights in space, would be an impediment to the development of space for both human habitation and commercial development. In 2015 the US passed legislation[16] stating that US citizens engaged in commercial recovery of an asteroid resource or a space resource, shall be entitled to any asteroid resource or space resource obtained, including to possess, own, transport, use, and sell the asteroid resource or space resource obtained in accordance with applicable law, including the international obligations of the United States.

## Outlook

Space science is likely to continue to advance further, as mankind seeks to push out and exploit outer space. National space efforts will be driven by strategic as well as economic interests. The high cost of space related activities will make governments play a dominant role and push them towards international cooperation wherever possible and beneficial. Permanent establishments in space and on nearby extraterrestrial bodies may become more feasible as technology advances. Unmanned space vehicles and robots will play a greater role. These developments will require diplomacy to resolve the issues that may arise, and prevent conflicts. In the event that contact is established with extraterrestrial life forms, a new and great challenge to diplomacy will arise.

The recent discoveries of an increasing number of exoplanets (outside the solar system) including some which appear to be capable of supporting life as we know it, opens up exciting possibilities for deep space exploration and colonization in the 21st century. This will bring with it new issues and problems that will require solutions.

## Endnotes

1   Jules Henri Giffard, a French engineer and inventor, built the first full-size airship — 44 metres long and a capacity of 3,200 cubic metres. A small 2.2-kilowatt steam engine drove a three-bladed propeller. The engine weighed 250 pounds (113 kilograms) and needed a 100-pound (45.4 kilograms) boiler to fire it. The first flight of Giffard's steam-

powered airship took place Sept. 24, 1852 at a speed of 10 kilometres/ hour,covering about 27 kilometers near Paris. See http://www.space. com/16623-first-powered-airship.html , accessed 29-3-2017

2   Sir George Cayley inscribed a silver medallion in 1799, that clearly depicted the forces that apply in flight. On the other side of the medallion Cayley sketched his design for a monoplane gliding machine.In 1804 Cayley designed and built a model monoplane glider of strikingly modern appearance. The model featured an adjustable cruciform tail, a kite-shaped wing mounted at a high angle of incidence and a moveable weight to alter the center of gravity. It was probably the first gliding device to make significant flights. See http://www.flyingmachines.org/ cayl.html   , accessed 29-3-2017

3   German V-1 (Vengeance-1) deployed in 1944 was the worlds first cruise missile powered by a pulsed jet engine and a mechanical autopilot. It had a weight of 2.2 tonnes and was 8.3 metres long, carried a 850 kg explosive warhead over an effective range of 250 km, with a speed of 600 km per hour, flying at about 700 metres altitude. It had an accuracy of 11 kms. The V-2  was the worlds first long range ballistic missile, deployed in 1944. It had a weight of 12.5 tonnes, length of 14 metres, carrying a 1 tonne explosive warhead over an effective range of 320 km. It used liquid fuel and had a speed of up to 5700 km per hour, and could reach a height of up to 200 km and was the first manmade object to reach space.

4   About ICAO, http://www.icao.int/about-icao/Pages/default.aspx   , accessed 2-5-2016

5   Spaceflight: The Development of Science, Surveillance, and Commerce in Space, R.D.Launius et al, Proceedings of the IEEE, 100, 1785, May 2012, http://ieeexplore.ieee.org/ielx5/5/6259910/06174432.pdf?tp=& arnumber=6174432&isnumber=6259910 accessed 12 Aug 2016

6   Space Law Treaties and Principles, UNOOSA, http://www.unoosa.org/ oosa/en/ourwork/spacelaw/treaties.html , accessed 17-8-2016

7   Space debris and human spacecraft, NASA, http://www.nasa.gov/ mission_pages/station/news/orbital_debris.html , accessed 17-8-2016

8  Changing Dynamics - India's aerospace industry, CII-PWC,2008, https://www.pwc.in/assets/pdfs/industries/changing-dynamics-india-aerospace-industry-091211.pdf , accessed 25-8-2016

9  ISRO successfully test-fires scramjet engine, The Hindu, 28-8-2016, http://www.thehindu.com/news/national/isro-successfully-testfires-scramjet-rocket-engine/article9042486.ece , accessed 29-8-2016

10 Indian-developed technology to boost range of missiles and protect re-entry vehicles news, Domain-b.com, 10 September 2008, http://www.domain-b.com/aero/mil_avi/miss_muni/20080910_Indian_technology.html ,accessed 25-8-2016

11 It May Irk China, But India Now Wants To Sell BrahMos Missile: Report, NDTV, 9 June 2016 http://www.ndtv.com/india-news/it-may-irk-china-but-india-now-wants-to-sell-brahmos-missile-report-1417070 , accessed 25-8-2016

12 History of Europe in Space, ESA website, http://www.esa.int/About_Us/Welcome_to_ESA/ESA_history/History_of_Europe_in_space , accessed 29-8-2016

13 International Cooperation, NASA, http://www.nasa.gov/mission_pages/station/cooperation/index.html , accessed 29-8-2016

14 About ISECG, ISECG website, http://www.globalspaceexploration.org/wordpress/?page_id=50 , accessed 29-8-2016

15 International Space Exploration Forum, http://iipdigital.usembassy.gov/st/english/texttrans/2014/01/20140110290302.html#axzz4IiFudDfb , accessed 29-8-2016

16 President Obama Signs Bill Recognizing Asteroid Resource Property Rights Into Law, Planetary Resources, 25-11-2015, http://www.planetaryresources.com/2015/11/president-obama-signs-bill-recognizing-asteroid-resource-property-rights-into-law, accessed 29-8-2016

# Chapter 9

# Ocean Space – the Frontier of the Depths

---

*"We must plant the sea and herd its animals using the sea as farmers instead of hunters. That is what civilization is all about - farming replacing hunting."*

— *Jacques Yves Cousteau, Oceanographer*

## Introduction

Man has for millennia been engaged closely with the oceans. Life itself sprang out of the benign ocean environment into the comparatively harsher environment of land. Oceans cover 71 percent of the surface area of our planet, and hold 97% of the planet's water. They produce more than half of the oxygen in the atmosphere, and absorb the most carbon from it. The importance of the oceans in global material and energy cycles is now beginning to be better appreciated, revealing the critical role of the oceans in atmospheric gas, energy and climate regulation, and for water, nutrient, and waste cycling. The oceans regulate global weather and climate[1] by providing a huge reservoir for heat energy and carbon dioxide. About half of the world's population lives within the coastal zone. The oceans have been estimated to contribute 21 trillion US$/year to human welfare (compared with a global GNP of 25 trillion US$). Over 90 million tonnes of fish are caught each year, representing an important food resource. Around a third of the oil and gas extracted worldwide comes from offshore sources. This figure is likely to continue to rise over the coming decades, for abundant oil and gas deposits still exist deep in the oceans. Ocean borne transport links the continents together and carries a 90 percent of global trade, with the shipping business amounting to some $ 380 billion per year, and is set to increase in future, because shipping is the most fuel efficient and carbon friendly form of commercial transport. The oceans represent an open access resource because of their fluid interconnectedness, their vast size, and the resulting difficulty of enforcing property rights to any particular area or resource.

With increasing human population and economic activity, the oceans have also become an area for competing interests of nations. In recent years, international agreements have come into existence seeking to reconcile competing and conflicting interests, carving out the ocean space into zones with different rights for nations. The world today has to deal with the threat posed by marine pollution, overexploitation of fish, and other human activities to the ocean ecosystem. It is not surprising therefore, that ocean space has become important for diplomacy and international relations.

## The Law of the Sea

Historically, the 'freedom of the seas' concept, dating from the 17th century restricted national rights to a sea area extending from a nation's coastlines, usually 3 miles (Three-mile limit), according to the 'cannon shot' rule developed by the Dutch jurist Cornelius van Bynkershoek. All waters beyond national boundaries were considered international waters free to all nations, but belonging to none of them. In the early 20th century, some nations expressed their desire to extend national claims so as to include mineral resources, to protect fish stocks, and to enforce pollution controls. In 1945 the US extended control to all the natural resources of its continental shelf. Other nations were quick to follow suit. Between 1946 and 1950, Chile, Peru, and Ecuador extended their rights to a distance of 200 nautical miles (370 km) to cover their fishing grounds. Other nations extended their territorial seas to 12 nautical miles (22 km).

By 1967, 66 nations had set a 12-nautical-mile (22 km) territorial limit, and eight had set a 200-nautical-mile (370 km) limit.

In 1956, the United Nations held its first Conference on the Law of the Sea (UNCLOS I) at Geneva, Switzerland. UNCLOS I resulted in four treaties concluded in 1958, regarding the Territorial Sea and Contiguous Zone (entry into force in 1964), the Continental Shelf (1964), the High Seas (1962) and on Fishing and Conservation of Living Resources of the High Seas (1966). It however failed to address the issue of width of territorial waters. In 1960, UNCLOS II conference was held but failed to produce any new agreements. With the passage of time the unresolved issue of varying claims of territorial waters became increasingly important. In 1973 the UNCLOS-III was convened, but in order to reduce the possibility of groups of nation-states dominating the negotiations, it used a consensus process rather than majority vote. With more than 160 nations participating, the conference lasted until 1982 and produced the Law of the Sea Convention (UNCLOS), which ultimately entered into force in 1994.

The most significant issues addressed in UNCLOS were the setting of territorial limits, navigation rights, archipelagic status and transit regimes, exclusive economic zones (EEZs), continental shelf jurisdiction, deep seabed mining, the exploitation regime, and protection of the marine environment, scientific research, and settlement of disputes. The convention set the limit of various areas, measured from a carefully defined baseline.

In order of decreasing sovereignty, UNCLOS defined several[2] zones described below–

(1) Internal waters covering all water and waterways on the landward side of the baseline. The coastal state is free to set laws, regulate use, and use any resource. Foreign vessels have no right of passage within internal waters.

(2) Territorial waters going up to 12 nautical miles (22 kilometres; 14 miles) from the baseline, where the coastal state is free to set laws, regulate use, and use any resource. Vessels were given the right of innocent passage through any territorial waters, with strategic straits allowing the passage of military craft as transit passage, in that naval vessels are allowed to maintain postures that would be illegal in territorial waters. "Innocent passage" is defined by the convention as passing through waters in an expeditious and continuous manner, which is not "prejudicial to the peace, good order or the security" of the coastal state. Fishing, polluting, weapons practice, and spying are not "innocent", and submarines and other underwater vehicles are required to navigate on the surface and to show their flag. Nations can also temporarily suspend innocent passage in specific areas of their territorial seas, if doing so is essential for the protection of its security.

(3) Archipelagic waters, defined in Part IV of UNCLOS, which also defines how the state can draw its territorial borders. A baseline is drawn between the outermost points of the outermost islands, subject to these points being sufficiently close to one another. All waters inside this baseline are designated Archipelagic Waters. The state has sovereignty over these waters (like internal waters), but subject to existing rights including traditional fishing rights of immediately adjacent states. Foreign vessels have right of innocent passage through archipelagic waters (like territorial waters).

(4) Contiguous zone, which extends beyond the 12-nautical-mile (22 km) limit, to a further 12 nautical miles (22 km) from the territorial sea baseline limit, in which a state can continue to enforce laws in four specific areas: customs, taxation, immigration and pollution, if the infringement started within the state's territory or territorial waters, or if this infringement is about to occur within the state's territory or territorial waters. This makes the contiguous zone effectively a "hot pursuit" area.

(5) Exclusive economic zones (EEZs), which extend out to 200 nautical miles (370 kilometres; 230 miles) from the baseline. Within this area, the coastal nation has sole exploitation rights over all natural (including living) resources. The EEZs were introduced to halt the increasingly heated clashes over fishing rights, although oil was also becoming important. The success of an offshore oil platform in the Gulf of Mexico in 1947 was soon repeated elsewhere in the world, and by 1970 it was technically feasible to operate in waters 4,000 metres deep. Foreign nations have the freedom of navigation and overflight, subject to the regulation of the coastal states. Foreign states may also lay submarine pipes and cables.

(6) Continental shelf defined as the natural prolongation of the land territory to the continental margin's outer edge, or 200 nautical miles (370 km) from the coastal state's baseline, whichever is greater. A state's continental shelf may exceed 200 nautical miles (370 km) until the natural prolongation ends. However, it may never exceed 350 nautical miles (650 kilometres; 400 miles) from the baseline; or it may never exceed 100 nautical miles (190 kilometres; 120 miles) beyond the 2,500-meter isobath (the line connecting the depth of 2,500 meters). Coastal states have the right to harvest mineral and non-living material in the subsoil of its continental shelf, to the exclusion of others. Coastal states also have exclusive control over living resources "attached" to the continental shelf, but not to creatures living in the water column beyond the exclusive economic zone. The precise limits of the continental shelf were left to be defined by a UN Commission on the Limits of the Continental Shelf (UNCLCS), based on claims submitted by interested states.

Aside from its provisions defining ocean boundaries, UNCLOS establishes general obligations for safeguarding the marine environment and protecting freedom of scientific research on the high seas, and also creates an innovative legal regime for controlling mineral resource exploitation in deep seabed areas beyond national jurisdiction, through an International Seabed Authority and the Common heritage of mankind principle. Landlocked states are given a right of access to and from the sea, without taxation of traffic through transit states. UNCLOS also provided a procedure for dispute settlement. Disputes can be submitted to the International Tribunal for the Law of the Sea established under the Convention, to the International Court of Justice, or to arbitration. Conciliation is also available and, in certain circumstances, submission to it would be compulsory. The Tribunal has exclusive jurisdiction over deep seabed mining disputes.

UNCLOS was the culmination of 9 years of negotiations over complex issues and a large number of nations with often conflicting interests that had to be reconciled. The issues connected to the regime for mining on the deep sea bed were particularly contentious, with US objections to it from 1982 to 1990. Negotiations between signatory and non-signatory states which took place during 1990-94 resulted in an agreement to sign a legally binding Convention. However, US interests were accommodated by stipulating that some key articles, including those on limitation of seabed production and mandatory technology transfer, would not be applied, that the United States, if it became a member, would be guaranteed a seat on the Council of the International Seabed Authority, and finally, that voting would be done in groups, with each group able to block decisions on substantive matters. The 1994 Agreement also established a Finance Committee that would originate the financial decisions of the Authority, to which the largest donors would automatically be members and in which decisions would be made by consensus.

The Law of the Sea is an effort to accommodate the conflicting interest of all countries in ocean space. These competing interests arose as technology advanced relating to the exploitation of ocean resources, especially oil, gas, and mineral resources on the sea floor. These were new issues added on to the older one of exploitation of fish and living resources in the oceans, something that had already resulted in conflicts between nations. A recent example if the South China Seas dispute, in which Chinese claims have been contested by several other countries. China's claims and aggressive postures have been linked to promises of large oil and gas resources in the disputed zone. Similarly the East China seas dispute is also driven by hydrocarbon resources. These disputes have acquired prominence and intensity today because technology for oil and gas exploitation has advanced to the point of being commercially viable. If technology related to deep sea bed mining and market conditions make such operations commercially viable in the future, we can expect to see a resurgence of differences over this subject.

## The Arctic

The Arctic consists of land, internal waters, territorial seas, exclusive economic zones (EEZs) and high seas. All land, internal waters, territorial seas and EEZs in the Arctic are under the jurisdiction of one of the five Arctic coastal states: Canada, Norway, Russia, Denmark (via Greenland), and the United States. The high seas including the North Pole and the region of the Arctic Ocean surrounding it are not owned by any country. The five surrounding Arctic countries are limited to an exclusive economic zone (EEZ) of 200 nautical miles (370 km; 230 mi) adjacent to their coasts. The waters beyond the EEZs of the coastal states are considered the "high seas" (i.e. international waters). The sea bottom beyond the exclusive economic zones and confirmed extended continental shelf claims are considered to be the "heritage of all mankind" where exploration and exploitation of mineral resources is administered by the UN International Seabed Authority. Norway, Russia, Canada, and Denmark launched projects to provide a basis for seabed claims on extended continental shelves to the UNCLCS, going beyond their exclusive economic zones. The United States has signed, but not yet ratified the UNCLOS.

The status of certain portions of the Arctic sea region is in dispute for various reasons. Canada, Denmark, Norway, Russia, and the United States all regard parts of the Arctic seas as national waters (territorial waters out to 12 nautical miles (22 km)) or internal waters. There also are disputes regarding what passages constitute international seaways and rights to passage along them. There is one single disputed piece of land in the Arctic—Hans Island—which is disputed between Canada and Denmark because of its location in the middle of an international strait.

Russia has claimed a large portion of the Arctic within its sector, extending to but not beyond the geographic North Pole on the basis that the Lomonosov Ridge, an underwater mountain ridge passing near the Pole, and Mendeleev Ridge on the Russian side of the Pole are extensions of the Eurasian continent. Russia has submitted new data reinforcing Russia's claim to part of the sea bottom beyond the 200-mile zone within its entire Arctic sector, the North Pole area included. On August 9, 2016 the UN Commission on the Limits of the Continental Shelf (UNCLCS) started working on the issue. The potential value of the North Pole and the surrounding area resides in new shipping routes and potential for large petroleum and natural gas reserves below the sea floor. However, outside undisputed EEZs, there is only a small unclaimed area of the Arctic potentially available for open gas/oil exploration.

## Management of Fisheries and Living Resources

The sustainable management of living resources of the oceans has been a particularly difficult problem. The fishing industry involves the livelihood of fishermen, food processors, and consumers on one hand. On the other hand, technology advances have increased exploitation to the point of depletion of fish stocks in the oceans. Fish can and do migrate freely throughout ocean space, irrespective of national jurisdictions. Therefore international cooperation and agreement has become a necessity for sustainable management of fish resources especially of highly migratory species such as Tuna. In such case, the cooperating nations need to agree on the level of fish catch and have a system of regulating fishing operators. Traditional fishing is a dying occupation, much as traditional hunting and gathering on land gave way to settled agriculture and animal husbandry. A similar transition in fishing is taking place, as aquaculture becomes more profitable and safe, and now accounts for half[3] of human fish consumption and an increasing share of employment, from 17 percent in 1990 to 33 percent in 2014. Technology can play a vital role in making aquaculture more productive and thereby provide a good alternative to traditional fishing, as well as offering high potential for women's employment.

UNCLOS specifies that fisheries should be managed sustainably, that fish resources should be utilized in an optimal manner, and that states, where necessary, should cooperate in this regard. Increased fisheries activities at the high seas during the 1980's and 1990's necessitated the negotiation of an additional treaty for improved control of fisheries at the high seas. The 1995 UN Fish Stocks Agreement is a key instrument for strengthening the management of fisheries resources based on the application of a precautionary approach to fisheries management, stronger rules for enforcement of regulations, more emphasis on regional cooperation, and better mechanisms for dispute resolution.

The FAO Code of Conduct for Responsible Fisheries is a non-binding agreement that gives guidelines for best practices for various aspects of fisheries and aquaculture. The Code includes a binding agreement aiming to enhance states control over vessels flying their flag and four International Plans of Action to remedy overcapacity in fishing fleets, illegal unreported and unregulated fishing and by-catches. The Code is also includes the Fish Code program for assistance to developing countries. Illegal fishing is a global threat to sustainable fisheries and to marine biodiversity. To combat this threat, the Agreement on Port State Measures against illegal fishing was adopted in 2009.

Under UNCLOS, an informal consultative process on the law of the sea (ICP) has addressed a range of issues relating to the sustainable use of the oceans. Fisheries-related issues that have recently been addressed in General Assembly resolutions include bottom-trawling at the high seas, the protection of vulnerable ecosystems, and application of ecosystem approaches to the management of ocean resources. FAO developed in 2008 the International Guidelines for the Management of Deep-sea Fisheries in the High Seas.

Fisheries have become an important issue in a number of global environmental discussions. At Rio+20 in 2012 the right to food and the role of healthy marine ecosystems, sustainable fisheries and sustainable aquaculture for food security and nutrition was highlighted. The precautionary approach, the ecosystem approach, and the integrated oceans management approach are responses to problems of increasing use and pressure on the resources of the oceans and the habitats and ecosystems.

## Large Marine Ecosystems and Marine Protected Areas

Large Marine Ecosystems (LMEs)[4] are large areas of ocean space of approximately 200,000 km² or greater, near coastal waters where living marine resources are more abundant and diverse than in open ocean areas. LMEs produce about 80% of the annual world's marine fisheries catch. However they are under great stress due to coastal ocean pollution and nutrient over enrichment, habitat degradation (e.g. sea grasses, corals, mangroves), overfishing, biodiversity loss, and climate change effects. This endangers the $12.6 trillion contributed annually by the 64 internationally recognized LMEs to the world's economy, and threatens the livelihood of millions of people dependent on marine resources.

The physical extent of the LME and its boundaries are based on four linked ecological, rather than political or economic, criteria. These are: (i) bathymetry, (ii) hydrography, (iii) productivity, and (iv) trophic relationships. Based on the 4 ecological criteria, 64 distinct LMEs have been delineated around the coastal margins of the Atlantic, Pacific and Indian Oceans. There is a 5-module strategy for measuring the changing status of LMEs, focused on indicators for measuring LME (i) productivity and oceanography, (ii) fish and fisheries, (iii) pollution and ecosystem health, (iv) socioeconomics and (v) governance. The LME concept is consistent with the UN Convention on the Law of the Sea and receives project based support from the UNEP and the Global Environment Facility (GEF).

Obviously, intergovernmental cooperation especially among the coastal states involved in each LME is crucial to the success of the LME approach to safeguarding marine biodiversity. In a number of cases political disputes among the coastal states can hamper such cooperation. This poses a challenge to policy makers and civil society which must find ways of securing effective cooperation among coastal states. Experience indicates that strong collaborative regional seas programmes, complemented by integrated coastal management by coastal states, is an important factor in successful management of LMEs. India has an important role to play being a major coastal state for two LMEs- the Arabian Sea (LME No. 32) and the Bay of Bengal (LME No. 34). Despite the great importance of these LMEs for India's economy, there has been insufficient public discussion over policy and strategy to manage these LMEs effectively.

The Arabian Sea LME No. 32 has an area of 3.95 million sq km[5] and the coastal states involved are Bahrain, Djibouti, India, Iran, Iraq, Kuwait, Oman, Pakistan, Qatar, Saudi Arabia, Somalia, United Arab Emirates, and Yemen. The coastal area stretches over 513 873 km2 with a population of 28 million in 2010, projected to increase to 110 million by 2100. This LME faces serious threats from overfishing, and marine pollution and is considered in a high risk situation among the LMEs. Despite these threats, the coastal states have not fashioned any effective cooperative arrangements to preserve this ecosystem. Lack of public awareness and political will, compounded by political differences among states such as India and Pakistan further compounds the problem.

The Bay of Bengal LME No. 34 has an area[6] of 3,66 million sq km and is bordered by Bangladesh, India, Indonesia, Malaysia, Maldives, Myanmar, Sri Lanka, and Thailand. The coastal area of 874 413 sq km has a population of 324 million thousand in

2010 projected to increase to 502 million in 2100. This LME is assessed as being at very high risk, due to problems of overfishing, and marine pollution, with very poor quality of data on important indicators. Coastal states have made some efforts at cooperation but there is no integrated LME management agency. There are two Regional Programme initiatives and several transboundary fisheries arrangements only one of which, the BOBLME project, which was launched by the FAO and may partially fill this role. It involves Bangladesh, India, Indonesia, Malaysia, Maldives, Myanmar, Sri Lanka, and Thailand and is focused largely on fisheries.

The South Asian Seas Action Plan (SASAP) launched in March 1995 involving Bangladesh, India, Maldives, Pakistan and Sri Lanka seeks to protect and manage the marine environment and related coastal ecosystems of the region covering areas which fall in LME 32 and LME 34. The South Asia Cooperative Environment Programme (SACEP) is acting as the Action Plan secretariat.

## Marine Protected Areas (MPAs)

Marine protected areas (MPA)[7] are protected areas of seas, oceans, estuaries or large lakes where human activity is limited for conservation purposes. Such marine resources are protected by local, state, territorial, native, regional, national, or international authorities and differ substantially among and between nations. This variation includes different limitations on development, fishing practices, fishing seasons and catch limits, moorings and bans on removing or disrupting marine life. In some situations, MPAs also provide revenue for countries, as an alternative to income from fisheries. The largest marine park in the world encompassing 1.55 million sq km was established by Australia in 2016 in the Ross Sea. Other large MPAs are in the Indian, Pacific, and Atlantic Oceans in certain exclusive economic zones of Australia and overseas territories of France, the United Kingdom and the United States. As of August 2016 there are more than 13,650 MPAs, encompassing 2.07% of the world's oceans, with half of that area – encompassing 1.03% of the world's oceans – receiving complete "no-take" designation (no fishing allowed)

MPAs may restrict fishing, oil and gas mining and/or tourism. Some fishing restrictions include "no-take" zones, which means that no fishing is allowed. Ship transit can also be restricted or banned, either as a preventive measure or to avoid direct disturbance to individual species. The degree to which environmental regulations affect shipping varies according to whether MPAs are located in territorial waters, exclusive economic zones, or the high seas. The law of the sea regulates these limits. Most MPAs have been located in territorial waters, where the appropriate government can enforce them. However, MPAs have been established in exclusive economic zones and in international waters. For example, Italy, France and Monaco in 1999 jointly established the Pelagos Sanctuary for Mediterranean, which covers both national and international waters.

India has 25 designated Marine Protected Areas in peninsular India[8], and an additional 106 MPAs located around its island territories in the Andaman, Nicobar, and Lakshadweep islands. The largest of the MPAs are located in West Bengal. India has also identified 12 protected areas as trans-boundary protected areas under the framework of the IUCN Transboundary Protected Area programme. Two of these sites are MPAs (Sundarbans National Park and Gulf of Mannar Biosphere Reserve).

Given the increasing attention being focused on the need to protect and preserve marine biodiversity, it is very likely that in future more and more MPAs will be established. This will require reconciling of the interests of various stakeholders and countries through diplomatic engagement.

## Oil and Gas resources

With the steady depletion of oil and gas onshore and the rise in oil and gas prices, there has been a shift towards exploration and exploitation of hydrocarbon resources in the sea floors. Starting from the near onshore shallow waters, technology has advanced to permit deep sea oil drilling. Drilling and production platform started off in 6 metres of water in 1947 and moved into deeper waters of up to 30 metres using fixed platform rigs and later using jack up rigs for depths up to 120 metres in the Gulf of Mexico. The first semi-submersible rig, an anchored, stable floating deep-sea platform Ocean Driller was launched in 1963 enabling deeper waters to be exploited. There are today over 620 mobile offshore drilling rigs. The world's deepest operation is in 2,600 meters of water.

Offshore oil and gas production presents many technology challenges due to the remote and harsher environment, stabilizing of floating deep sea rigs, and the pressure of sea water. Some of the production operations are carried out subsea, by separating water from oil and re-injecting it rather than pumping it up to a platform, or by flowing to onshore, with no installations visible above the sea. Subsea installations enable operations at progressively deeper waters and overcome challenges posed by sea ice such as in Arctic regions. The development of directional drilling in the 1970s enabled oil and gas to be extracted from as much as 10 km laterally, and facilitated near onshore operations.

Offshore oil production involves substantial environmental risks, most notably oil spills from oil tankers or pipelines transporting oil from the platform to onshore facilities, and from leaks and accidents on the platform. Produced water is also generated, which is brought to the surface along with the oil and gas; it is usually highly saline and may include dissolved or unseparated hydrocarbons. In April 2010, the highly advanced state of the art Deepwater Horizon platform, an ultra-deepwater, dynamically positioned, semi-submersible offshore oil drilling rig 80 km off-shore of the US coast exploded, and sank two days later. The resulting undersea oil leak took 5 months to stop, and released some 700,000 tonnes of oil making it the worst oil spill in US history. The full range of available technologies were used to mitigate the damage caused by the oil discharge, and the legal proceedings cost British Petroleum $

54 billion for cleaning up and a further $ 19 billion for settlement of legal cases. The incident also led to review and changes in policy and regulation of off shore oil and gas drilling. Given the large number of ongoing offshore drilling operations, there is a risk that another accident could occur possibly impacting more than one coastal state, and this would require international action involving governments as well as private business. Demand for greater safety and regulation of offshore drilling have already appeared and there may be moves to define a set of international standards in this least internationally regulated marine activity sector. Some ideas that have been proposed[9] are - (1) developing and strengthening regional agreements on the environmental safety of offshore oil and activities; (2) the elaboration of an international convention regulating liability and compensation for pollution damage resulting from offshore drilling activities; (3) Building of States' capacities in effectively controlling the offshore industry.

The oceans also contain natural gas hydrates which occur on continental margins and shelves worldwide from polar regions to the tropics, and their energy content is estimated to exceed that of all other fuel sources combined. In the deep seabed, vast amounts of methane are produced by the decay of organic matter and as a by-product of microbial activity. As the methane rises towards the seafloor, it reaches a zone known as the Gas Hydrate Stability Zone (GHSZ), where the pressures and temperatures are right for the methane to become trapped as gas hydrate. Gas hydrate usually appears as a white, ice-like substance that often forms lumpy crystals or layers within the sediment. Natural gas hydrate resources are being exploited by Germany, USA, Japan, South Korea, India, Taiwan and China. The potential hazards associated with these new technologies have so far been largely ignored. US Geological Survey estimates that global gas hydrates contain between 10,000 trillion cubic feet to more than 100,000 trillion cubic feet of natural gas representing more carbon trapped inside hydrates than is contained in all known reserves of fossil fuels. If methane hydrates begin to be exploited on a large scale, it could present a serious climate change challenge.

## Deep Sea bed Mining

Interest in metalliferous deposits on the seafloor as a valuable source[10] of scare metals has become increasingly widespread in the past 5 years, though the concept of deep-sea mining was first introduced in the 1960's. Over the past decade the demand for precious metals for advancing technologies has rocketed, making deep-sea deposits increasingly attractive to commercial operators. The most likely targets for deep-sea mining are polymetallic sulphides, manganese nodules and cobalt-rich ferromanganese crusts. On a longer time scale, rare earth elements (REEs) in deep-sea muds may also become important. Mineral exploration activities have already taken place in prospective areas of the ocean beyond national jurisdiction under license from the International Seabed Authority, most notably around the Clarion-Clipperton Zone in the Pacific Ocean, parts of the Indian Ocean and along the Mid-Atlantic Ridge. In territorial waters, commercial activity has progressed more rapidly and the extraction of gold, copper and silver from deep water deposits offshore Papua New Guinea is close to becoming a reality.

Polymetallic massive sulphide deposits are most commonly formed along tectonic plate boundaries and volcanic provinces in water depths from 500 to 5000 metres . Deposits of this type containing copper, lead, zinc, precious and trace metals range from several thousand to several million tonnes, and it is estimated that around 600 million tonnes of massive sulphide deposits occur within the easily accessible neovolcanic zone of mid-ocean ridges. Mining of copper and gold from sulphides was due to commence of Papua New Guinea in 2014, but has been delayed. Recently China, the Russian Federation, France and South Korea have been granted exploration licenses by the ISA for massive sulphides. New license applications have been submitted to the International Seabed Authority to carry out exploration for polymetallic sulphides in the central Indian Ocean.

Polymetallic nodules (also commonly called manganese nodules) most abundant on abyssal plains at water depths of 4000-

6500 metres, are probably the most likely commodity to be developed into a commercial operation. They contain commercially attractive (though variable) levels of metals such as nickel, copper, cobalt, and traces of other valuable metals such as molybdenum, zirconium and REEs. They extraction is relatively easy when compared to some other metal deposits in the deep sea. The nodules of greatest commercial interest occur in the Clarion-Clipperton Zone in the equatorial Pacific Ocean (CCZ) and in the Central Indian Ocean Basin. In the CCZ, polymetallic nodules cover 9 million km2 with typical concentrations of 15 kg m-2. The nodules contain nickel, copper, and cobalt (around 2 - 3% of the nodule weight) as well as traces of other metals such as molybdenum, Rare Earth Elements and lithium, which are important to high-tech industries. The amount of copper contained in the CCZ nodules is estimated to be about 20% of that held in global land-based reserves.

Rare Earth Elements (REEs) are key components of many new digital and e-mobility technologies. Demand for them has reached 120,000 tonnes in 2010 - exceeding the world's current annual production of 112,000 tonnes. In 2011, it was reported that high concentrations of REEs (in particular the rarer heavy REEs) could be found in the top few metres of deep-sea clays[11]. Although it is not clear whether the resource is commercially viable, deep-sea mud mining is being explored.

The extraction of deep-sea mineral resources will have a significant impact on the marine environment, particularly its ecosystems. Deep-sea mining will potentially affect extensive areas of seabed, could lead to emission of toxic materials into the ocean, or the discharge of fine particulate material. There is therefore an urgent need to assess the nature and scales of the potential impacts of mining, and how they will affect deep-sea ecosystems. This is an area where international cooperation to evolve effective regulations and standards at an early stage is essential.

## Technology and Ocean space issues

The future success of nations in ocean space will depend on the availability of modern ships, undersea vehicles, aircraft, satellites, laboratories, and observing systems, as well as the continuous development and integration of new technologies into these facilities[12]. Modern facilities and new technologies that can operate in the open ocean, along the coasts, in polar regions, on the seafloor, and even in space will become critical. Technology development will cover land-based structures (such as laboratories and monitoring stations) as well as remote platforms (such as ships, airplanes, satellites, and submersibles) for conducting research, observations, monitoring, and enforcement activities are conducted; hardware such as research equipment, instrumentation, sensors, and information technology systems; and support services including expert human and software resources needed to operate and maintain the facilities and hardware and carry out activities in ocean space.

An example is ocean energy, where tidal and wave energy production in the EU is projected to rise to 66 MW by 2018. This technology is ready for commercial application. This "Blue Energy Action" by the EU could create jobs and produce renewable energy on a large scale. Other coastal states could develop similar programmes, leading to reduction in use of fossil fuels and consequent climate change benefits. Underwater drones[13] and other submersible vehicles are an active area of development. Another development is of wave gliders recently purchased by Japan, which are long-duration, environmentally friendly ocean robots that collect and communicate real-time ocean data without using fuel.

The earth's oceans contain dissolved materials of considerable commercial value, such as magnesium, lithium, and uranium, which could be exploited if it can be extracted. US Scientists have developed a material[14] that can effectively pull uranium (concentration of 3.3 parts per billion) out of seawater. A total of 4 billion tons of uranium is available in all Earth's seawater. Magnesium is a valuable metal used for making light weight alloys and its main source is seawater and briny lakes. Each cubic kilometre of seawater contains more than a million tons of magnesium compounds as well as over 45 minerals and metals[15].

As technology for ocean resource exploitation develops, and ocean based economy takes off (the so called "blue economy") there will be much greater advantages for coastal states versus land locked states. The divide between these two groups will resurface as it did during UNCLOS negotiations. Land locked states would be free to exploit resources in the deep sea zones, and may demand greater transit and connectivity rights to these zones.

Technology may well lead to greater militarization of ocean space. Underwater drones could make surface warships more vulnerable to attacks. Submarines with nuclear armed missiles are considered essential for providing strategic nuclear deterrence. However technology may lead to easier detection and means of neutralizing these submarines in times of conflict, leading to less certainty of deterrent capability. Another factor could be the existence of maritime disputes which may lead to increase of assertiveness by some states, for example in the case of the South China Seas dispute. Undersea cables and pipelines could also be damaged for example by drones in the event of conflicts. In the case of the Arctic Ocean, there are unresolved disputes over certain areas, which could lead to tensions. These may intensify as technology makes exploration and exploitation of the Arctic more attractive in future.

However some technologies could bring nations together. Ocean energy and sea water mining offer possibilities for stronger international collaboration. Technology for Carbon Dioxide sequestration in the oceans is another possibility. Products for health care and wellness derived from marine living resources may be another. The UN has adopted Sustainable Development Goal 14 which seeks by 2030 "to Conserve and sustainably use the oceans, seas and marine resources for sustainable development". This will be a challenging task to accomplish.

# Endnotes

1   The ecological, economic, and social importance of the oceans, Robert
    Costanza, Ecological Economics 31, 1999, p199–213, https://www.pdx.
    edu/sites/www.pdx.edu.sustainability/files/Costanza%20Oceans%20
    1999.pdf , accessed 1-9-2016

2   United Nations Convention on the Law of the Sea of 10 December
    1982 Overview and full text, UN Division for Ocean Affairs, http://
    www.un.org/depts/los/convention_agreements/convention_overview_
    convention.htm , accessed 1-9-2016

3   State of the World's Fisheries and Aquaculture 2016, FAO, http://www.
    fao.org/3/a-i5555e.pdf, accessed 2-9-2016

4   Large Marine Ecosystems of the world, NOAA, http://www.lme.noaa.
    gov/index.php?option=com_content&view=article&id=1&Itemid=11
    2 , accessed 27-1-2017

5   LME 32 Arabian Sea fact sheet, onesharedocean.org, http://
    onesharedocean.org/public_store/lmes_factsheets/factsheet_32_
    Arabian_Sea.pdf , accessed 27-1-2017

6   LME 34 Bay of Bengal fact sheet, onesharedocean.org, http://
    onesharedocean.org/public_store/lmes_factsheets/factsheet_34_Bay_
    of_Bengal.pdf , accessed 27-1-2017

7   What is a marine protected area ? Ocean Service, NOAA, http://
    occanscrvice.noaa.gov/facts/mpa.html , accessed 27-1-2017

8   Coastal and Marine Protected Areas in India: Challenges and Way
    Forward, Sivakumar, K, August 2014, researchgate.net, https://www.
    researchgate.net/publication/265642777_Coastal_and_Marine_
    Protected_Areas_in_India_Challenges_and_Way_Forward , accessed
    27-1-2017

9   International regulation of offshore oil and gas activities: time to head
    over the parapet, IDDRI, 2014, http://www.iddri.org/Publications/

Collections/Syntheses/PB0614_JR_offshore_EN.pdf , accessed 3-9-2016

10 The deep sea as a target for exploitation, MIDAS, http://www.eu-midas. net/science , accessed 3-9-2016

11 Discovery of rare earth metals in ocean mud could help Japan, autoblog. com, 4 April 2013, http://www.autoblog.com/2013/04/04/discovery-rare-earth-metals-ocean-mud-japan , accessed 3-9-2016

12 An Ocean Blueprint for the 21st Century Final Report of the U.S. Commission on Ocean Policy, Chapter 27, http://govinfo.library.unt. edu/oceancommission/documents/full_color_rpt/27_chapter27.pdf , accessed 2-9-2016

13 The Military Is Working To Develop Underwater Drones, Taskand purpose.com, 22 August 2016, http://taskandpurpose.com/the-military-is-working-to-develop-underwater-drones , accessed 2-9-2016

14 Uranium Extraction from Seawater Takes a Major Step Forward, Scientific American, 1 July 2016, http://www.scientificamerican.com/ article/uranium-extraction-from-seawater-takes-a-major-step-forward , accessed 2-9-2016

15 Over 40 minerals and metals contained in seawater, their extraction likely to increase in the future, Mining Weekly, 1 April 2016, http://www. miningweekly.com/article/over-40-minerals-and-metals-contained-in-seawater-their-extraction-likely-to-increase-in-the-future-2016-04-01/ rep_id:3650 , accessed 2-9-2016

# Chapter 10

# Nanotechnology – Technology of the Small with Huge Possibilities

*"And it turns out that all of the information that man has carefully accumulated in all the books in the world can be written in this form in a cube of material one two-hundredth of an inch wide – which is the barest piece of dust that can be made out by the human eye. So there is plenty of room at the bottom!"*

*– Richard P Feynman, 1959*

## Introduction

Nanotechnology is a rapidly expanding and multidisciplinary field of science and technology. It may be defined as the science, engineering, and technology related to the understanding and control of matter[1] at the length scale of approximately 1 to 100 nanometres (nm)[2]. One nm is one billionth of a metre, and for purposes of comparison, the width of an average hair is 100,000 nm. Visible light has a wavelength of between 400 and 600 nm. Human blood cells are 2,000 to 5,000 nm long, a strand of DNA has a diameter of 2.5 nm, and a line of ten hydrogen atoms is one nm long.

Nanotechnology can be seen as a stage of mankind's ability to understand and manipulate matter at ever smaller scales as time goes by. Over the last century, scientists have developed a much more detailed understanding of matter at finer and finer levels. At the same time, engineers have gradually acquired the ability to reliably manipulate material to increasingly finer degrees of precision. Although we have much knowledge about individual atoms, much remains to be known about the ability to actually see them, manipulate them, and understand how they behave in groups at the nano-scale level.

Nanostructures can be nano-scale in one dimension as in the case of thin films or sheets. They could be nano-scale in two dimensions as in the case of carbon nanotubes and DNA, the building block of life. They can be nano-scale in three dimensions, as in the case of nanoparticles. There are also nanoporous structures, where the cavities are nano-sized, while the bulk material remains macro sized. Examples are nanoporous membranes, and nanosponges. The discovery of nanostructures with just one element Carbon, which was recognized by Nobel prizes in 1996[3] and in 2010[4] with novel properties illustrates the rich variety of nanostructures that remain to be explored involving combinations of known elements.

Nanotechnology is much more than simply the scaling down of behaviour of micro and macro levels of matter. The basic building blocks of matter and life occur at the nano level. Molecular chemistry, genetic reproduction, cellular processes, and the frontiers of electronics occur on the nano level. Understanding how these processes work and being able to manipulate events at this level in order to get specific outcomes, opens up the possibility of significant new advances in a wide variety of fields including electronics, medicine, and material sciences. Nanotechnology holds out the promise of new tools, products, and technologies to address global challenges[5] such as clean, secure and affordable energy; stronger, lighter and more durable materials; low cost devices to provide clean drinking water; medical devices and drugs to detect and treat diseases more effectively with less side effects; lighting with much lower energy consumption; sensors to detect and identify harmful chemical or biological agents; and techniques to remove hazardous chemicals from the environment.

The small size of nanostructures can significantly change the properties of bulk matter. Nanostructures have relatively much larger surface area, so surface properties may dominate and make them much more reactive. At the nano level quantum physics effects become significant and the properties of materials can differ considerably from the properties of either individual atoms or bulk matter. Material at the nanoscale can exhibit surprising characteristics that are not evident at large scales. For example, collections of gold nanoparticles can appear orange, purple, red, or greenish[6], depending upon the specific size of the particles making up the sample. Carbon atoms in the form of a carbon nanotube can exhibit tensile strengths 100 times that of steel and can be either metallic or semiconducting depending on their configuration. Titanium dioxide and zinc oxide, common ingredients in sun screen, both appear white when made of macro particles, but when the particles are ground to the nanoscale, they appear translucent.

## Nanotech and Development, and Nano-divides

Nanotechnology certainly has the potential to enable mankind to overcome numerous challenges of development. Examples are creation of cheap water filters that do not require electricity, solar cells that can be printed on sheets of paper, and improved drug delivery systems for neglected diseases. Besides direct impact on the world's poor, there are also several other development benefits. Production of nanomaterials could create new jobs; nanotechnology applications may help to improve infrastructures for water and electricity; high-skilled jobs in nanotechnology could be created in developing countries, nanotechnology products using locally available minerals could boost domestic value addition; and nanotechnology could stimulate economic growth.

The term "nano-divide" is used to refer to concerns that nanotechnology may cause a widening of the gap between the rich and poor with benefits accruing largely to the richer parts of the world[7]. There have been concerns about the impact of nanotechnology such as intellectual property rights, displacement of jobs , and trade agreements involving nano-products. Several kinds of nano-divides have been identified. These are access nano-divide, where the poor do not have access to beneficial nano-technologies; profit nano-divide, where economic gains from nano technologies may flow largely to the developed countries, exacerbating the gap with developing countries; benefit nano-divide, where nanotechnologies may be developed primarily to meet the needs of wealthy consumers, to the detriment of addressing the needs of the poor; and control nano-divide, where the control over nanotechnologies may be used to dictate the terms of their use thereby deepening unequal power balances. It is clear that market forces alone would result in accentuating several of these divides, unless policy correctives are present.

Experts have identified nanotechnology areas most relevant for development, and have suggested that benefits to developing countries were especially expected in the fields of energy (storage, production and conversion), agricultural productivity enhancement, and water treatment and remediation, and disease diagnosis and screening. These could be the basis for identifying nanotechnology areas that would best help the achievement of the UN's Sustainable Development Goals 2030. However, efforts to assess the impact of nanotechnology on development should include the views of various other experts besides scientists and engineers so as to reflect the technological and societal changes required for nanotechnology to make an impact on global poverty. Rather than considering developing countries merely as passive recipients of nanotechnologies and products, importance should also be given to capacity and capability building in nanotechnology in developing countries.

## Major Countries and Nanotech Initiatives, Global Spending Trends

Total global spending on nanotechnology research and development (R&D) by governments, corporations, and venture capital (VC) investors is estimated at $18.1 billion in 2014, with $5.9 billion of this spending from the U.S. Government. The U.S. leads in government (state and federal) nanotechnology funding with $1.72 billion spent in 2013 and $1.67 billion spent in 2014. Europe's collective spending (European Commission and individual country programs) was $2.45 billion in 2014, an increase of 9.8% from 2012. Next are Japan and Russia at $1.1 billion each, and China and South Korea at about $500 million each. The next largest spenders were Canada, Taiwan, Brazil, Singapore, Israel, and India. While some countries, such as the U.S., have centralized government programs to coordinate nanotechnology activities, many countries no longer explicitly fund nanotechnology, though it may be a part of initiatives that are funded under different technology support programs.

The U.S. and Japan continue to serve as role models for the rest of world when it comes to corporate spending on nanotechnology, contributing $4.0 billion and $2.4 billion, respectively, in 2014. U.S.-based companies received the vast majority of VC investments, capturing 72% (a slight increase from 70% in 2012) of the $316 million invested in companies developing nanotechnology in 2014. Revenues from nano-enabled products grew to $1.61 trillion in 2014, from $848 billion in 2012. Total sales of final products that incorporate emerging nanotech in some fashion grew from $850 billion in 2012 to $1.6 trillion in 2014, an increase of 90%. The global value of nanomaterials increased 35% over 2012 to reach $2.12 billion in 2014.

The US launched the National Nanotechnology Initiative in 2000, establishing a multi-agency program to coordinate and expand federal efforts to advance the state of nanoscale science, engineering, and technology, and to position the United States to lead the world in nanotechnology research, development, and commercialization. Spending has been approximately $22.3 billion for the NNI from 2001 to 2016 and is projected at $1.443 billion for 2017 almost the same as in 2016. There are 5 programme focus areas which also include 5 specific signature initiatives.

In terms of scientific articles output, China has the most publications (with approximately 39,500 publications in 2014), followed by the European Union 27 (EU-27, approximately 33,500), the United States (approximately 24,000), South Korea (approximately 8,000), and Japan (approximately 7,000). An analysis of nanotechnology related patents from 1986 to 2015 showed that the United States had the highest share of patents (24.4%) followed by China(17.7%) of the patents; Japan (10.3%); South Korea (7.9%); Germany (4.2%); Taiwan (2.3%); Russia (2.0%); France (1.7%); the United Kingdom (1.4%); and Canada (1.0%). These figures indicate roughly the ranking of various countries in nanotechnology R & D.

## Nanotech and Strategic and Economic Competitiveness

Like other branches of science and technology, nanotechnology also holds out the promise of more efficient and more innovative products that can meet a variety of needs. This can translate into significant strategic and economic gains, including global competitiveness. This is the reason why nanotechnology continues to attract funding from both public and private sources across the globe.

The strategic importance of Nanotechnology[8] arises from the potential to influence warfare technology in large number of ways. Lighter, stronger, heat resistant nanomaterial could be used in producing all kinds of weapons, making military transportation faster, strengthening armour and saving energy. Specialized composite nanomaterial coatings with nonreflecting properties may be able to confer stealth capabilities to platforms. Nanosensors have a large variety of military applications such as detection of nuclear, biological and chemical agents, battlefield surveillance, monitoring of equipment health. Other applications are of nano sized communication devices and nano device enabling of military equipment. These are only a few illustrations of the tremendous potential of nanotechnology in the military field.

It will become quite important to be able to detect very small devices—perhaps even sub-microscopic devices. Massive arrays of sensors could determine the outcome of conflicts. It may prove impossible to protect civilians and civilian property from nano weapons. A nano-arms race could be unstable. Nuclear weapons are hard to design and build, require easily monitored testing, do indiscriminate and lasting damage, do not rapidly become obsolete, have almost no peaceful use, and are universally abhorred. On the other hand, Nano capability and nano weapons will be easy to build, to test undetected, will be relatively easy to control and deactivate. They would become obsolete very rapidly, almost every design is dual-use, and peaceful and non-lethal use will be common, and barriers to deployment and use would be minimal.

In the civilian field, nanotechnology has undoubtedly great potential. New materials could result in new products and upgrading of older products to add new features. Molecular level assembly could result in scalable more efficient processes for producing useful chemicals, transforming the chemical industry in ways that are hard to imagine. Nanotechnology enabled smart production lines could make production flexible and  dispersed rather than concentrated, making logistics far cheaper.

Some of the possibilities that nanotechnology could bring include the synthesis of nano-materials and nano-components with corresponding technical applications in areas such as green energy, electronics, catalysis, LEDs, sensors, composites and biomedicine. Nanotechnology currently underpins many practical applications and has the potential to further enhance quality of life and environmental protection. Applications can be found everywhere, such as in clothing, cars, windows, computers, displays, cosmetics and medicine with new functionalities, intelligence, portability, and networking capability in many new products with high market potential.

For example, nanomanufacturing will increasingly allow mass reproducibility at an extremely precise scale and could open new world markets by making low cost goods similar in function to existing products. There are potential applications across a huge range of sectors, from improved battery-powered vehicles to more targeted medical therapies to nanotube-enhanced road pavement with remote sensing capabilities[9]. Nanotechnology researchers have also improved food safety and biosecurity, produced lightweight but strong nanocomposites for building more fuel-efficient vehicles, created methods for separating carbon dioxide from other gases, and dramatically improved the efficiency of plastic solar cells.

## Nanotech and Environment Health and Safety

Nanomaterials have been the subject of concern from the environmental and health point of view. In particular, the release of nanoparticles into the environment and their impact on living beings is the subject of considerable research. As nanoparticles are extremely reactive chemically and can easily move into and through biological systems, their toxicological effects would be correspondingly greater. Nanotechnology development can be viewed across four overlapping generations of new nanotechnology products and processes, each generation having its own unique characteristics and risks: passive nanostructures, active nanostructures, complex nanosystems, and molecular nanosystems. The extent of knowledge and ability to control nanostructure behaviour is greatest in the first generation, while the potential social and ethical impacts are more transformative in the case of the other three generations.

There is a need for greater understanding of the potential risks of nanomaterials and further studies are required for : (1) hazard characterization, in areas such as toxicity, ecotoxicity, carcinogenicity, volatility, flammability, persistence and accumulation in cells; and (2) exposure, including the potential for nanomaterials to enter the body through oral, cutaneous and inhalation during production, transport (in air, water, soil and biosystems), decomposition or waste disposal activities.

Several studies have brought out some human health risks. Large doses of nanoparticles can cause cells and organs to demonstrate a toxic response even when the material itself is non-toxic. Some nanoparticles are able to penetrate the liver and other organs and to pass along nerve axons into the brain. Nanomaterials can combine with iron or other metals, thereby increasing the level of toxicity and presenting unknown risks. Engineered nanomaterials could have unknown characteristics, properties, and potential impact in concentrated amounts. Nanopowders have increased risk of dust explosion and the ease of ignition. Nanostructures with the potential for bioaccumulation, including absorption of contaminants such as pesticides, metals and organic compounds could then move along the food chain. Non-biodegradable nanostructures may persist in the environment with unknown effects.

An example is the case of air pollution in cities, arising from burning of fuels and vehicles, where small sized particles are emitted into the atmosphere and accumulate and persist for long periods. Particles of 2.5 microns and below (PM2.5 or 2500 nanometres) are absorbed easily into the blood stream and penetrate deep into the lungs, and cause diseases[10]. Smaller sized particles remain in the air longer, spread over longer distances, and are more easily absorbed into the body. The impact of PM1 particles (1000 nanometres) in urban air is insufficiently known, but there are some indications that it could be an even greater threat than PM2.5. PM0.1 (ultrafine particles (UFP)), those particles with a diameter less than 100 nm (including nanoparticles (NP)) are considered especially dangerous to human health[11] and may contribute significantly to the development of numerous respiratory and cardiovascular diseases such as chronic obstructive pulmonary disease (COPD) and atherosclerosis.

Decisions taken about the direction and level of nanotechnology research and development (R&D) may affect investment in key areas to benefit future economic development. It might cause uneven distribution of nanotechnology risks and benefits among different countries and economic groups. Nanotechnology might be used for criminal or terrorist activity. There could be a race for military applications. If the knowledge within professional communities is not appropriately shared with regulatory agencies, civil society and the public, then risk perception would not be based on the best available knowledge, and incorrect decisions might be taken.

The risks faced by any individual, company, region or country depend not only on their own choices but also on those of others. The transboundary effects of nanotechnology including its risks may cause international disputes and tensions.

## Regulation and Nanotechnology

The risks associated with nanotechnology require appropriate regulatory systems at the national level, as well as effective international cooperation in setting standards and harmonizing regulatory practices in order to enable international trade in nano related products. This is likely to be a serious challenge, as not enough information is available about the various risks associated with nanotechnology products and processes. The approach of modifying the existing regulatory systems for food, cosmetics, etc to include nanotechnology products might not be adequate. Also, at present there is insufficient coordination on nanotechnology safety issues between the different actors and stakeholders, particularly those in science, industry, consumers, government regulators, civil society and international bodies. For example, there is a gap between regulatory provisions, their areas of relevance and different standards applied to the same product. Regulatory practices and guidelines for nanotechnology products are in their infancy and much development lies in the future.

## International Engagement in Nanotechnology

International co-operation is essential for the development of nanotechnology, to meet scientific and technical challenges and to better focus nanotechnology research and overcoming knowledge gaps more rapidly. Synergies can be created that can contribute to improve the quality of life in all parts of the world. International co-operation in nanotechnology is needed both with countries that are economically and industrially advanced (to share knowledge and profit from critical mass) and with those less advanced (to secure their access to knowledge and avoid any 'nano divide' or knowledge apartheid). An international dialogue on a responsible development and use of nanotechnology is needed, as well.

## Nanotechnology and India

The potential of nanotechnology was realized by government of India as early as in the year 2001 when Nanoscience and Technology Initiative (NSTI) was launched as a mission mode program in the 10th Five Year Plan (2002-2007) with a budget of Rs 65 million led by the Department of Science and Technology (DST). The NSTI was followed by Nano Mission (2007–12) with a budget allocation of Rs. 10 billion[12]. Nano Mission is an umbrella program for capacity building which envisages the overall development of the field of research in the country and to tap some of its applied potential for the nation's development. Nano Mission Phase II (2012-17) was approved with an allocation of Rs 65 billion. However the decrease in government spending on nanotechnology since 2007 is a source of concern, given the considerable potential of nanotechnology for India's economic development. Recognising the need for safety, draft guidelines and best practices[13] were issued in 2016 by the Government of India for research laboratories and industries.

## Conclusions and Future Projections.

Nanotechnology has been recognized as a cross cutting transformative technology. It could result in major developments such as new materials, new manufacturing processes, sensors, and other devices. Along with these changes there could be new threats and challenges, These could have a major impact on society and international relations in the future. Many countries and businesses are investing large resources into research and development in nanotechnology and the results could impact on their international competitiveness. There is the possibility of an emerging nano-divide similar to the digital divide within and among societies. There is insufficient knowledge of the risks of nanotechnology even among experts, and therefore greater awareness and education of policy makers and civil society is important. International cooperation in this field is therefore a necessity in order to meet challenges and derive full benefits. Much of this architecture remains to be constructed.

## Endnotes

1 Small Wonders, Endless Frontiers: A Review of the National Nanotechnology Initiative (2002), page 5, https://www.nap.edu/read/10395/chapter/3 accessed 11-10-2016

2 The upper end is not strictly defined, and may extend up to 500 nm in some cases.

3 Harold Kroto, Roert Curl, and Richard Smalley, awarded the 1996 Nobel Prize in Chemistry for discovery of fullerenes in 1985

4 Andrei Geim and Konstatntin Novoselov ,awarded the 2010 Nobel Prize in Physics for the discovery of graphene.

5 Nanotechnology - Big things from a tiny world, National Nanotechnology Initiative USA, 2008, http://www.nano.gov/sites/default/files/pub_resource/nanotechnology_bigthingsfromatinyworld-print.pdf , accessed 11-10-2016

6 Nanotechnology: A Gentle Introduction to the Next Big Idea,Mark and David Ratner, Prentice Hall Professional, 2003. Much of the color in the stained glass windows found in medieval and Victorian churches and some of the glazes found in ancient pottery depend on the fact that nanoscale properties of materials are different from macroscale properties. In particular, nanoscale gold particles can be orange, purple, red, or greenish, depending on their size.

7 Broadening nanotechnology's impact on development, Koen Beumer, Nature Nanotechnology, 11, 398 (2016)

8  Nanotechnology: The Emerging Field for Future Military Applications, Sanjiv Tomar, IDSA Monograph Series No.48, October 2015, http://www.idsa.in/system/files/monograph/monograph48.pdf ,accessed 15-10-2016

9  Economic Growth and Breakthrough Innovations:A Case Study of Nanotechnology, Lisa Larrimore Ouellette, Economic Research Working Paper No. 29, WIPO November 2015, http://www.wipo.int/edocs/pubdocs/en/wipo_pub_econstat_wp_29.pdf , accessed 24-10-2016

10 Toxic air pollution particles found in human brains, The Guardian, 5 Sept 2016, https://www.theguardian.com/environment/2016/sep/05/toxic-air-pollution-particles-found-in-human-brains-links-alzheimers , accessed 17-4-2017

11 Inhaled Pollutants: The Molecular Scene behind Respiratory and Systemic Diseases Associated with Ultrafine Particulate Matter, , H Traboulsi et al, International Journal of Molecular Sciences, 2017, 18, 243, http://www.mdpi.com/1422-0067/18/2/243/pdf , accessed 17-4-2017

12 Mission on Nano Science and Technology (Nano Mission), Department of Science and Technology, Government of India, http://nanomission.gov.in , accessed 15-10-2016

13 Draft Guidelines and Best Practices for Safe Handling of Nanomaterials in Research Laboratories and Industries, Department of Science and Technology, Government of India, http://nanomission.gov.in/What_new/Draft_Guidelines_and_Best_Practices.pdf , accessed 17-4-2017

# Chapter 11

# Climate Change and Energy – Still time to Save our Planet

*"I'd put my money on the sun and solar energy. What a source of power! I hope we don't have to wait till oil and coal run out before we tackle that."*

*– Thomas Alva Edison, American inventor, scientist, and businessman, 1931*

## Introduction

In the period since the industrial revolution, human activity has resulted in large amounts of gases being released into the atmosphere. The impact of these emissions has been manifold. The emission of ozone depleting substances degrades the ozone layer, which shields the surface of the earth from harmful ultraviolet radiation. Recognizing this, countries have collectively agreed on measures to phase out the use of ozone depleting substances through instruments such as the Montreal Protocol. On the other hand, emissions of other gases such as Carbon Dioxide, etc, known as greenhouse gases, increases the greenhouse effect of the atmosphere, resulting in increase in net energy absorbed from the sun, and thereby causing global warming and climate changes. The greenhouse gas emissions have accumulated in the atmosphere during the period when industrial countries developed and also at present when the developing countries are developing their economies. According to experts, the total cumulative amount of greenhouse gases in the atmosphere must be kept within certain limits if global temperature rise is to be kept within a limit of 2 degrees Celsius. Beyond this point, large scale irreversible changes in the earth's climate could cause severe impacts and pose threats to human survival. Limits on greenhouse gases if they are to be accomplished without stifling development of developing countries, would require major advances in technology and possibly changes in life styles. This has placed climate change at the forefront of international negotiations over the past two decades, without reaching a final solution. Climate change is an area where there is close interplay between technology, society and politics, requiring adequate understanding of all these areas. Within and among nations, the price of controlling climate change has forced discussions over the hard choices to be made as mankind travels collectively on its journey on the space station that we call earth.

## Evolution of the Atmosphere and Climate

The Earth's climate is regulated by a complex balance of factors. The driving force is solar radiation which is partially absorbed and reflected into space. The fraction that is absorbed depends in a delicate way on the interplay of the atmosphere and the Earth's surface, and which is also responsible for the Earth's climate and weather systems. The amount of solar energy received depends on the sun's energy output and distance from the sun, while the rotation of the earth about its axis affects the distribution of the received energy over its surface. These factors are common to all the planets of the solar system. What is unique to the Earth is the composition of the atmosphere and the existence of life, which affect each other.

Solar output and the Earth's atmospheric composition have undergone major changes over geologic time scales going back to approximately 4.5 billion years ago, when the Earth and the solar system are regarded as being formed. At that time solar output was 30 % lower than at present, while the atmosphere was composed of over 30 percent of carbon dioxide with no oxygen. Free oxygen in the atmosphere was produced by anaerobic cyanobacteria (blue-green algae), by organic photosynthesis of carbon dioxide ($CO_2$) and water ($H_2O$) a process that releases oxygen as a by-product. By about 1 billion years ago, the atmosphere had approximately reached the present composition with 21 % oxygen and Carbon Dioxide had fallen to below 1 percent. The increase in solar output over this period was compensated partly by the reduced greenhouse effect due to decrease in $CO_2$ levels from nearly 38 percent to 0.04 percent. As the sun ages, its output will continue to increase further [1]. At the end of the next 4.8 billion years, the Sun will be about 67% brighter than it is now. In the 1.6 billion years following that, the Sun's luminosity will rise to a 220 % of the present value. The Earth by then will have been roasted to bare rock, its oceans and all its life boiled away by an enlarged sun that will be some 60% larger than at present.

The above indicates that the Earth's atmosphere and climate have undergone major changes over geologic time scales. However, we are more concerned with the changes over the span of the past 20 thousand years, during which human civilization has come into existence. Over such short term period, the sun's output remains steady with fluctuations of less than 0.15% while it undergoes 11 and 22 year cycles. The above facts do illustrate that human existence is so tiny a fraction of the solar system's life span which has witnessed so many dramatic changes.

## The Greenhouse effect

To understand the dynamics of climate change, one must understand the greenhouse effect. In very simplified terms, the earth's atmosphere functions like a greenhouse, letting in solar energy to the earth's surface, but not allowing the energy radiated by the earth's surface to escape into space. Solar radiation reaching the earth is mostly concentrated in the visible region of the spectrum, (400-600 nanometres wavelength) which passes without absorption through the atmosphere. On the other hand, the earth's surface emits radiation mostly in the long wavelength infra red region of the spectrum (4000 - 10000 nanometres), much like a hot kettle. This radiation is absorbed by the atmosphere, and does not escape into space. The absorption of infrared radiation by the atmosphere is critical to the greenhouse effect, which arises from the additional radiation of energy from the atmosphere back to the earth's surface. The actual energy balance of the energy received from the sun is shown in figure 11.1

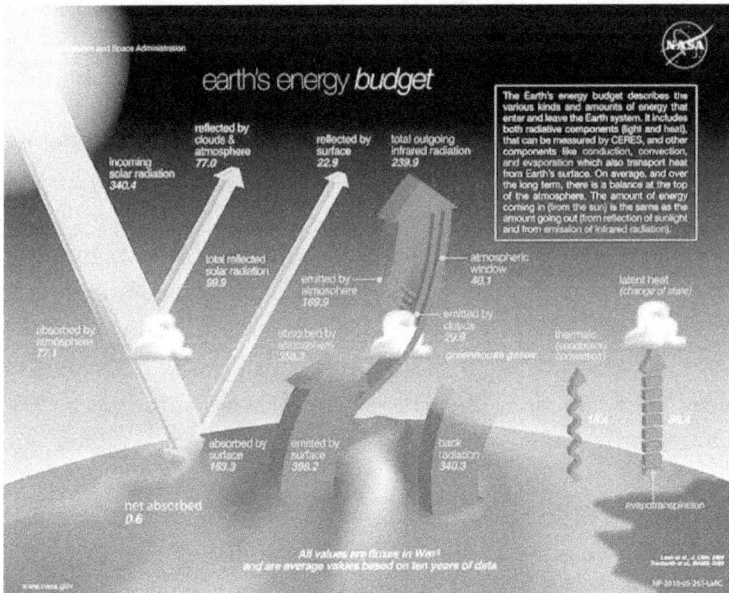

Figure 11.1 The Earth's energy budget (NASA)

The major constituents of the atmosphere[2] (Nitrogen, 78 percent) and Oxygen (21 percent) do not absorb the infra red radiation from the earth's surface. The most important infrared absorbers that remain for long periods in the atmosphere and contribute to the greenhouse effect are Carbon Dioxide (0.04 percent), Water vapour (variable 0 to 5 percent), and other gases such as Methane (1.8 parts per million by volume, ppmv), Ozone (found in the upper layers of the atmosphere), Nitrous Oxide and Chlorofluorocarbons. If the earth had no atmosphere and was a perfect black body, its temperature would be -14 degrees Celsius, while the actual average temperature of the earth is 33 degrees higher, due to the greenhouse effect.

The actual atmosphere is not homogenous, its composition, pressure, density and temperature varies with height. Therefore each layer of the atmosphere has to be analyzed in detail for its contribution to the greenhouse effect. The process of energy transfer from the earth's surface through the atmospheric layers containing absorbing gases is known as radiative transfer, similar to the way that energy from the sun's interior diffuses outward. The infrared radiation emitted from the upper levels of the earth's atmosphere (approx 5-6 kilometres height, where the atmosphere is thin) can escape into space. The earth's surface is also not uniform, with seas, ice cover, and land having different properties and receiving varying amounts of solar energy that depend on time of day, latitude and season. This forms part of global climate modeling using elaborate computer calculations to simulate the actual earth's behaviour. In terms of impact, these calculations have shown that the greenhouse effect contributions of various components of the atmosphere[3] are Water vapour (36-70 percent), Carbon Dioxide (9-26 percent), Methane (4-9 percent), and Ozone (3 -7 percent). The presence of clouds which absorb and emit infrared radiation also affects the situation.

The impact of a greenhouse gas depends on how much infrared radiation it can absorb, and how long it remains in the atmosphere (its lifetime). These are translated in to the term "global warming potential" (GWP) which measures its global warming impact relative to Carbon Dioxide over various time spans. Carbon Dioxide has a lifetime of between 100 to 300 years in the atmosphere, depending on uptake by the oceans and plants. Methane has a lifetime of 12 years and GWP of 86 over 20 years and 29 over 100 years, while Nitrous Oxide has a lifetime of 114 years and GWP of 289 over

20 years and 290 over 100 years. Sulphur Hexafluoride, used in the electrical industry has a very long lifetime of 3200 years and very high GWP of 16300 over 20 years and 22300 over 100 years and could be particularly harmful. International discussions have often taken the GWP 100 years into account, but this would underestimate the impact of Methane which has a much higher GWP over 20 years, and which is likely to be released in greater quantities as a result of the growth[4] in natural gas usage and hydraulic fracturing (fracking).

Climate change scientists use the term radiative forcing (RF) for the rate of energy change per unit area of the globe as measured at the top of the atmosphere. The climate sensitivity specifically due to carbon dioxide is often expressed as the temperature change in degrees Celsius associated with a doubling of the concentration of carbon dioxide in Earth's atmosphere. Carbon dioxide dominates the total forcing, with methane and chlorofluorocarbons (CFC) becoming relatively smaller contributors to the total forcing over time. The five major greenhouse gases account for about 96% of the direct radiative forcing by long-lived greenhouse gas increases since 1750. The remaining 4% is contributed by the 15 minor halogenated gases. Relative to 1750, the total RF had reached 2.936 Watts per sq metre by 2014. The best estimate of the climate sensitivity of the earth is 0.8 degrees Celsius per Watt per sq metre, which indicates that the global average temperature rise since 1750 would be about 2.4 degrees Celsius. It is the total RF from all greenhouse gases combined (with carbon dioxide being the most important) that needs to be limited in order to limit global warming.

## Man-made Atmospheric Changes and Climate Changes

Human civilization has evolved over the period going back some 20000 years. During this period the composition of the atmosphere has remained largely stable . Carbon dioxide levels have ranged between 180 to 280 ppmv over the past half million years, while the earth's climate passed through ice ages at intervals of 40 to 100 thousand years.

The present phase of global warming is attributed to increasing emissions of $CO_2$ and other greenhouse gases into the Earth's atmosphere The global annual mean concentration of $CO_2$ in the atmosphere has increased by more than 40% since the start of the Industrial Revolution, from 280 ppmv, the level it had for the last 10,000 years leading up to the mid-18th century, to 399 ppmv as of 2015. This corresponds to the large scale use of carbon bearing fossil fuels such as coal, oil and gas and deforestation. The present concentration is the highest in at least the past 800,000 years and probably the highest in the past 20 million years. The average concentration of atmospheric $CO_2$ is currently rising at a rate of approximately 2 ppmv/year and accelerating. An estimated 30–40% of the $CO_2$ released by humans into the atmosphere dissolves into oceans, rivers and lakes, which contributes to ocean acidification.

The exact impact of increased green house gas accumulations in the atmosphere on global climate has been the subject of much scientific study and especially with computer based global climate models for the earth's climate. There are also many feedback loops such as the effect of water vapour and oceans that are poorly understood. The present state of the science of global climate modeling enables only a rough estimate to be made of global temperature rise due to increase in greenhouse gas concentrations in the atmosphere. This has led to challenges from "climate change skeptics" who have questioned the basis of the argument that human activity can lead to climate change. However, the majority of the scientific community and especially the Intergovernmental Panel on Climate Change (IPCC) which represents the collective global expertise on this subject do agree that human activity according to the business as usual scenario will result in increased greenhouse gas emissions leading to global warming and climate change. It is important to heed the precautionary principle – the absence of scientific certainty should not rule out taking precautionary action against possible catastrophes.

## Ozone Layer changes and impact

The Ozone layer has been a source of some confusion in the public mind. While ozone does act as a greenhouse gas, its concentration in the global atmosphere is far too low to cause more than a 3 to 7 percent of the total greenhouse effect. Therefore we may ignore the impact of ozone depletion on climate change to a first approximation[5]. Of much greater importance is the beneficial effect of the ozone layer in shielding the earth's surface from harmful ultra violet radiation which can damage DNA and life. The loss of the ozone layer can result in increased exposure to harmful ultra violet radiation, and this is the main reason for concern over ozone layer depletion due to human activity.

The suns radiation contains significant amounts of short wavelength radiation (less than 400 nanometres). This ultraviolet radiation is harmful to living organisms and in fact is intentionally used in sterilizing equipment. Fortunately this radiation is absorbed in the upper layers of the atmosphere by nitrogen ($N_2$) and oxygen ($O_2$) with the latter forming ozone ($O_3$). The ozone layer contains less than 2-8 parts per million of ozone and is mainly found in the lower portion of the stratosphere, from approximately 20 to 40 kilometres above Earth, although its thickness varies seasonally and geographically. Most of the ozone is found in the mid-to-high latitudes of the northern and southern hemispheres, and the highest levels are found in the spring, and the lowest in the autumn in the northern hemisphere.

The ozone layer was discovered in 1913. Its properties were explored in detail by the British meteorologist G. M. B. Dobson, who developed instruments that could be used to measure stratospheric ozone from the ground. Between 1928 and 1958, Dobson established a worldwide network of ozone monitoring stations, which continue to operate to this day. The "Dobson unit", a measure of the amount of ozone in the total air column[6], is named in his honor.

Biologically harmful ultraviolet (UV) radiation coming from the sun has several components. Very short UV (10–100 nm) is absorbed by Nitrogen while UV-A (400–315 nm), UV-B (315–280 nm), and UV-C (280–100 nm) are not absorbed. UV-C, which is very harmful to all living things, is entirely screened out by a combination of dioxygen $(O_2)$ (< 200 nm) and ozone $(O_3)$ (> about 200 nm) by around 35 kilometres altitude. UV-B radiation can be harmful to the skin and is the main cause of sunburn; excessive exposure can also cause cataracts, immune system suppression, and genetic damage, resulting in problems such as skin cancer. The stratospheric ozone layer (which absorbs from about 200 nm to 310 nm with a maximal absorption at about 250 nm) is very effective at screening out UV-B. Nevertheless, some UV-B, particularly at its longest wavelengths, reaches the surface, and is important for the skin's production of vitamin D. Ozone is transparent to most UV-A, so most of this longer-wavelength UV radiation reaches the surface, and it constitutes most of the UV reaching the Earth. This type of UV radiation is significantly less harmful to DNA, although it may still potentially cause physical damage, premature aging of the skin, indirect genetic damage, and skin cancer.

Ozone can be destroyed by a number of free radical catalysts which have both natural and man-made sources. Human activity has dramatically increased the levels of chlorine (Cl) and bromine (Br) which are liberated from the parent compounds by the action of ultraviolet light. On average, a single chlorine atom is able to react with 100,000 ozone molecules before it is removed from the catalytic cycle. This fact plus the amount of chlorine released into the atmosphere yearly by chlorofluorocarbons (CFCs) and hydrochlorofluorocarbons (HCFCs) demonstrates how dangerous CFCs and HCFCs are to the environment.

The thickness of the ozone layer varies considerably worldwide, due to atmospheric circulation patterns as well as variations in solar intensity. It is smaller near the equator and larger towards the poles and thicker during the spring and thinner during the autumn. The lowest amounts of column ozone found anywhere in the world are over the Antarctic in the southern spring period of September and October and to a lesser extent over the Arctic in the northern spring period of March, April, and May. Ozone levels have dropped by a worldwide average of about 4 percent since the late 1970s. For approximately 5 percent of the Earth's surface, around the north and south poles, much larger seasonal declines have been seen, and are described as "ozone holes". The discovery of the annual depletion of ozone above the Antarctic was first announced in 1985. The depletion of the stratospheric ozone layer was found to be much greater in the temperate latitudes, resulting in increased exposure to ultraviolet radiation.

## International response to ozone layer depletion

During the 1970s, scientific studies indicated that chemicals produced by human activity could harm the stratospheric ozone layer, especially by releasing chlorine. Public concern grew over ozone layer depletion and its consequences for human health such as increased risk of skin cancer. In the USA, the project to develop a stratospheric supersonic transport aircraft lost support and was abandoned. The non essential use of CFCs as propellants for aerosols was banned in 1978[7]. Under public pressure governments drew up a World Plan of Action in 1977 on the Ozone Layer, which called for intensive international research and monitoring of the ozone layer. A global framework convention on stratospheric ozone protection the Vienna Convention, was concluded in 1985, in which States agree to cooperate in relevant research and scientific assessments of the ozone problem, to exchange information, and to adopt "appropriate measures" to prevent activities that harm the ozone layer. The obligations are general and contain no specific limits on chemicals that deplete the ozone layer. Consensus could not be reached between the EEC and the Toronto group of countries led by the US on specific reduction targets for certain chemicals[8]. In 1986 and 1987 a new sense of urgency about stratospheric ozone emerged as a result of rapid growth in demand for CFCs and the discovered Antarctic ozone hole, an event which received much media attention. The US pushed for a freeze and near complete phase out within 14 years of ozone depleting substances. European governments under pressure from environmental groups, as well as Japan, feared that the United States might take unilateral action and impose trade sanctions. An international agreement was also considered by all parties, including industry, as a powerful incentive for developing and marketing CFC substitutes. Negotiations resulted in the conclusion of the Montreal Protocol, which went into effect on January 1, 1989.

The Montreal Protocol outlines specific measures and timetables for reducing production and consumption of CFCs and halons. The protocol divides ozone-depleting compounds into two groups. Group I includes the fully halogenated CFCs (CFC-11, -12, -113, -114, and -115) which are the most threatening to the ozone layer, and Group II includes the halons. The protocol also makes an important distinction between developed countries and developing countries, which are referred to as "Article 5 countries" over the timing of the production and consumption reductions. This difference is based on the principle of "common but differentiated responsibility" of the developed and developing countries, based on the logic of greater historical responsibility for the release of damaging substances[9]. The former are required to do more to cut down and phase out ozone depleting substances, while the latter are given an extra ten years delay. In addition a Multilateral Fund[10] was set up under Article 10 of the Protocol to assist developing countries to meet their commitments under the Protocol. The Montreal Protocol has built in flexibility and includes mechanisms for re-evaluating and revising the protocol[11] on the basis of new scientific information beginning in 1990.

Has the Montreal Protocol achieved its objectives? According to an expert assessment conducted in 2014, the total concentration of all ozone depleting substances of human origin in the atmosphere after reaching a peak in 1993, declined by 10 percent by 2012. Some studies indicate that without the Montreal Protocol, the ozone layer thickness would have declined from around 300 DUs to below 100 DU by 2070. It has been estimated that about 2 million skin cancer cases per year may have been avoided due to the Montreal Protocol. These studies indicate that the Montreal Protocol has indeed been successful in preventing ozone layer degradation and reducing human health risks, and that the benefits would become more apparent in the future.

The important drivers behind the success of the negotiations to finalize the Montreal Protocol were - (1) the evolving scientific understanding of stratospheric ozone and its influence on policymaking; (2) increasing public concern based on the threat of skin cancer and the perception of potential global catastrophe associated with the discovery of the Antarctic ozone hole; and (3) the availability of acceptable substitutes and alternative technologies. Also the ozone layer depletion was most severe in the regions of the earth, in which most of the industrialized countries lay. The increased risk of skin cancer was seen by the public as a global problem which is threatening to future generations, increasing, hard to prevent, and not easily reduced.

The ozone agreements are the first to address a long-term problem in which the effects are not evident for decades after the cause. Decisions were taken on the basis of probabilities, since damage had not yet occurred. Since scientific understanding of the problem was not complete, the agreements needed to be flexible and capable of being adapted to accommodate new scientific assessments. No single country or group of countries could address the problem of ozone depletion alone, so maximum international cooperation was needed.

The Montreal Protocol reflects a convergence of interest of scientists who warned of growing threats to the ozone layer, private industry that wanted a level playing field as companies responded to new national legislation controlling the harmful chemicals, citizens groups and nongovernmental organizations advocating environmental protection, and national governments that increasingly saw an international agreement as in their own best interests. The Protocol has been regarded as one of the most successful international treaties concluded, in terms of its universality and achievement of objectives[12]. It is therefore an extremely useful model and precedent for future international environmental agreements, designed to preserve planet earth. Unfortunately the deviations from the basic principles underlying the Protocol have resulted in difficulties in reaching agreement on controlling global warming and climate change. In the case of the Montreal Protocol, the controls have been imposed on chemicals which have been deliberately produced for various applications, while in the case of global warming, the greenhouse gases are incidentally produced and released and not accounted for in the course of economic activities of a much more widespread nature. This remains a formidable technical challenge.

## Greenhouse Gas Changes due to Human Activity

The concentration of greenhouse gases in the earth's atmosphere has been steadily increasing[13] since the pre-industrial period (the 1750s). During the million years preceding the 1750s, the carbon dioxide concentration ranged between 180 - 300 parts per million (ppmv). The post 1750 increase to nearly 400 ppmv can largely be attributed to human activity, especially the industrial revolution that swept across the globe. The Table 11.1 below gives details of greenhouse gas concentrations compiled from various sources, up to April 2016.

Table 11.1

## Greenhouse gas concentrations in April 2016 compared to pre-industrial levels

| Gas | Pre-1750 tropospheric concentration | Recent tropospheric concentration | GWP (100-yr time horizon) | Atmospheric lifetime (years) | Increased radiative forcing (W/m$^2$) |
|---|---|---|---|---|---|
| Concentrations in parts per million (ppm) | | | | | |
| Carbon dioxide $(CO_2)$ | ~280 | 399.5 | 1 | ~ 100-300 | 1.94 |
| Concentrations in parts per billion (ppb) | | | | | |
| Methane $(CH_4)$ | 722 | 1834 | 28 | 12.4 | 0.50 |
| Nitrous oxide $(N_2O)$ | 270 | 328 | 265 | 121 | 0.20 |
| Tropospheric ozone $(O_3)$ | 237 | 337 | n.a. | hours-days | 0.40 |
| Concentrations in parts per trillion (ppt) | | | | | |
| CFC-11 $(CCl_3F)$ | zero | 232 | 4,660 | 45 | 0.060 |
| CFC-12 $(CCl_2F_2)$ | zero | 516 | 10,200 | 100 | 0.166 |

The table also shows the effect of increase in concentrations in terms of radiative forcing, which is the net increase in solar energy absorbed per square metre due to greenhouse gas increase. The effect of this is to contribute to an increase in global temperature[14]. Concentrations of ozone and water vapor vary over the earth and over time. The globally averaged water vapor concentration is about 5,000 ppm. A warmer atmosphere will likely contain more water vapor than at present. Of the various trace gases in the atmosphere, CFC-11 and CFC -12 are the largest contributors to global warming, and these have extremely long lifetimes, but contribute less due to their low concentration. Fortunately, international efforts to stop their emissions under the Montreal Protocol have resulted in stopping the increase in their concentration.

## Methane, Natural Gas exploitation

Apart from Carbon Dioxide, the next most important greenhouse gas is Methane (CH4) which is the major constituent of natural gas. Methane has a global warming potential over 100 years (GWP-100) of 29 times carbon dioxide. Methane has a half-life of 7 years in the atmosphere so its global warming potential over a 20-year time period (GWP-20) is 86 times that of carbon dioxide. The use of GWP-100 for methane could therefore underestimate its actual impact on global warming.

Pre 1750s, for about 800,000 years, the concentration of methane was fairly steady at below 722 parts per billion (ppb), but had risen to 1800 ppb by 2011. The growth rate of methane declined from 1983 until 1999. From 1999 to 2006, the atmospheric methane levels were fairly constant, at about 1780 parts per billion (ppb) but since 2007, globally averaged CH4 has been increasing again. Causes for the increase during 2007-2008 include unusual climatic conditions, and increase in production, usage, and distribution of natural gas across the globe. From 2014 to 2015 global methane increased substantially faster (11.5 ppb/yr) than it had from 2007 to 2013 (5.7 ± 1.2 ppb/year). Methane could be produced from rice production, melting of permafrost[15], and from the ocean gas hydrates. But the recent more rapid increase could probably be linked to greater exploitation of natural gas from unconventional sources due to advances in technology[16] including growth of hydraulic fracturing ("fracking") and shale gas exploitation. Fracking reportedly releases large amounts of natural gas – which consists of both $CO_2$ and methane – into the atmosphere. In fact, fracking wells leak 40 to 60 per cent more methane than conventional natural gas wells[17]. According to one study, the United States alone could be responsible for between 30-60% of the global growth in human-caused atmospheric methane emissions since 2002 because of a 30% spike in methane emissions[18] across the country. It would seem therefore that the advantages of natural gas over other fossil fuels as far as climate change impact may have to be reconsidered. Significantly US official sources tend to highlight the natural sources of methane rather than the sources related to the natural gas exploitation. The correlation between rise in methane emissions and fracking needs to be explored in more detail.

## Climate Change impact

The increase of greenhouse gases in the earth's atmosphere results in an increase in net solar energy absorbed by the earth. This has an impact on the earth's climate, in particular on the earth's global average temperature. The earth's climate is a complex phenomenon, affected by various factors and feedback loops. We do not have an adequate understanding of the climate of the earth. Global climate models (GCM) are an attempt to seek a better understanding. These are complex mathematical representations of the major climate system components (atmosphere, land surface, ocean, and sea ice), and their interactions and the energy balance between them. These models require large computational capacity. Climate models agree on certain basic aspects of future climate change. For example, they all show rising global temperatures with amplified warming in the Arctic, enhancement of the hydrologic cycle (dry places becoming dryer and wet places becoming wetter), and rising sea level. Many of these factors affect each other and could be drastically altered in an already changing climate. Climate models can disagree on many results and projections due to natural variability, differences in forcing, and differences in feedbacks. Forcings vary greatly among climate models and the main uncertainties arise from the impact of aerosols on earth's energy balance. Climate relationship to aerosols, such as the existence and concentration of water vapor and clouds, differs because of this uncertainty.

The most discussed aspect of climate change is the relationship between carbon dioxide concentrations and the rise in global average temperature. Connecting these two is a chain of cause and effect with considerable uncertainty at various steps. In general, the total radiative forcing effect of all greenhouse gases is first calculated. Then the climate models are used to predict the effect of this increase in forcing. It has been said that stabilization of carbon dioxide levels at 450 ppm would result in rise in temperature of over 2 degrees Celsius by 2100. There is considerable variation in the various climate models on the impact of radiative forcing effect. For example, models predict between 25 to 70 percent probability of exceeding 2 degrees Celsius by 2100 if carbon dioxide is stabilized at 450 ppm. At 500 ppm the probability ranges from 50 to 95 percent.

Measurements show that each of the last three decades has been successively warmer at the Earth's surface than any preceding decade since 1850. The period from 1983 to 2012 was likely the warmest 30-year period of the last 1400 years in the Northern Hemisphere. The globally averaged surface temperature show a warming of 0.85 [0.65 to 1.06] degrees Celsius over the period 1880 to 2012. Figures 11.2 and 11.3 show a striking parallel between the rise in global average temperature and the rise in greenhouse gas levels[19] over the period 1850-2014.

Figure 11.2. Global average temperature rise, 1850-2014

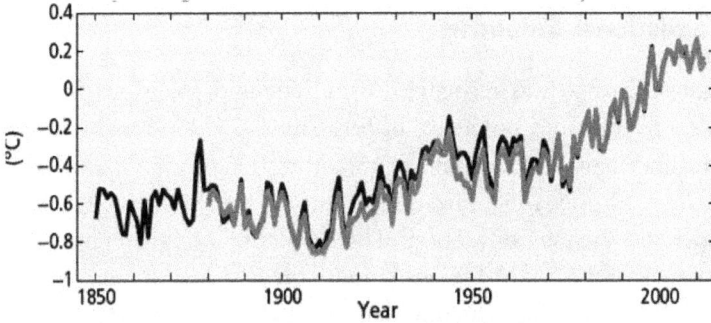

Figure 11.3. Globally averaged greenhouse gas concentrations
1850-2014

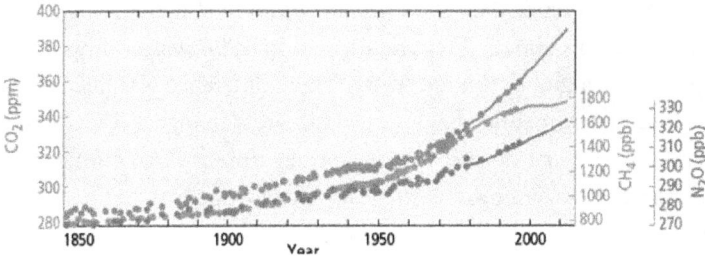

The conclusion emerging from the above is that there is evidence that human activity has caused rise in greenhouse gas levels in the atmosphere and that there has been significant global warming. The rise in global temperature is not uniform across the earth, it is much greater towards the Polar Regions. Global warming also increases the probability and severity of extreme climate events such as storms, cyclones, floods, and droughts. There may not be adequate scientific understanding of the quantity of global warming or the climate change impact attributable to greenhouse gas increases, but the absence of scientific certainty should not be a reason for not taking preventive action, the so-called "precautionary principle"[20]

## International Responses

Climate change was recognized as a serious problem at the first World V Climate Conference in February 1979. 1985 was the year that the Vienna Convention for the Protection of the Ozone Layer was created and 1987 saw the signing of the Montreal Protocol under the Vienna convention. This model of using a Framework conference followed by Protocols under the Framework was seen as a promising governing structure that could be used as a path towards a functional governance approach that could be used to tackle broad global warming.

In 1988 the Intergovernmental Panel on Climate Change (IPCC) was created by the World Meteorological Organization and the United Nations Environment Programme as an authoritative expert body to assess the risk of human-induced climate change. The Second World Climate Conference held in 1990 was an important step towards a global climate treaty. The Conference was held when the Intergovernmental Panel on Climate Change (IPCC) had completed its First Assessment Report in time for the conference, which in turn was to provide critical input for the first session of the International Negotiating Committee for a Framework Convention on Climate Change (INC). The resulting Ministerial Statement, however, did not offer a high level of commitment. The scientists and technology experts at the Conference issued a strong statement highlighting the risk of climate change and agreed that it was time for the world community to take strong measures to reduce sources and to increase "sinks" of greenhouse gases, despite the remaining scientific uncertainties. The Conference issued a Ministerial Declaration only after hard bargaining over a number of difficult issues and did not specify any internationally agreed targets.

The key agreed elements included in the Declaration, were (1) recognition of a number of principles including the concept of climate change as a common concern of humankind, the principle of equity and the common but differentiated responsibility of countries at different levels of development, the concept of sustainable development, and the precautionary principle; (2) the need for further scientific research on the causes and effects of climate change and recommend that this be done mainly through support of the World Climate Programme (WCP); (3) that response measures must be adopted without delay, despite remaining scientific uncertainties; (4) that developed states, which are responsible for 75% of the world's emissions of greenhouse gases, should establish targets and/or feasible national programmes or strategies which will have a significant effect on limiting emissions of greenhouse gases (4) that the emissions from developing countries must still grow to accommodate their development needs; nevertheless, these states should, with support from the developed nations and international organizations, take action; and (5) the call for elaboration of a framework treaty on climate change and the necessary protocols - containing real commitments and innovative solutions - in time for adoption by the UN Conference on Environment and Development (UNCED) in June 1992.

An Intergovernmental Negotiating Committee produced the text of the United Nations Framework Convention on Climate Change (UNFCCC) in May 1992 and was formally adopted the Earth Summit in Rio de Janeiro in June 1992. Strong pressure from civil society groups especially in the developed countries played a key role in ensuring success. The relative success of the Montreal Protocol also provided a precedent that could be emulated. The UNFCCC objective is to "stabilize greenhouse gas concentrations in the atmosphere at a level that would prevent dangerous anthropogenic interference with the climate system". The framework set no binding quantitative limits on greenhouse gas emissions for individual countries and contains no enforcement mechanisms. Instead, the framework provides for specific international treaties (called "protocols" or "Agreements") to set binding limits on greenhouse gases and deal with related questions such as monitoring, enforcement, finance and technology. UNFCCC entered into force on 21 March 1994, and has 197 parties as of December 2015. The convention enjoys broad legitimacy, due to its nearly universal membership.

Article 3(1) of the Convention states that Parties should act to protect the climate system on the basis of "common but differentiated responsibilities", and that developed country Parties (Annex I Parties) should "take the lead" in addressing climate change. Under Article 4, all Parties make general commitments to address climate change through, for example, climate change mitigation and adapting to the eventual impacts of climate change. Article 4(7) states that the extent to which developing country Parties will effectively implement their commitments under the Convention will depend on the effective implementation by developed country Parties of their commitments under the Convention related to financial resources and transfer of technology and will take fully into account that economic and social development and poverty eradication are the first and overriding priorities of the developing country Parties.

The parties to the UNFCC have met annually from 1995 in Conferences of the Parties (COPs) to assess progress in dealing with climate change. In 1997, the Kyoto Protocol was adopted and established legally binding obligations for developed countries (Annex I Parties) to reduce their greenhouse gas emissions in the period 2008-2012. It entered into force in 2005 and currently has 192 Parties. Resistance to binding cuts and to provide financial and technical support to developing countries was apparent in the attitude of several major developed countries. The US in particular did not ratify the Kyoto Protocol while Canada withdrew in 2012.

Under the Kyoto Protocol, 38 industrialized countries and the European Community (the European Union-15, made up of 15 states at the time of the Kyoto negotiations) committed themselves to binding targets for greenhouse gas emissions. The targets apply to the four greenhouse gases carbon dioxide ($CO_2$), methane ($CH_4$), nitrous oxide ($N_2O$), sulphur hexafluoride ($SF_6$), and two groups of gases, hydrofluorocarbons (HFCs) and perfluorocarbons (PFCs). These six greenhouse gases are translated into CO2 equivalents using their global warming potentials (GWPs) in determining reductions in emissions. These reduction targets are in addition to the industrial gases, chlorofluorocarbons, or CFCs, which are dealt with under the 1987 Montreal Protocol on Substances that Deplete the Ozone Layer.

Under Article 4.1 of the Protocol, only the Annex I Parties have committed themselves to national or joint reduction targets (formally called "quantified emission limitation and reduction objectives". The other Non Annex I Parties) are mostly developing countries, and may participate in the Kyoto Protocol through the Clean Development Mechanism. The emissions limitations of Annex I Parties vary considerably. Some Parties have emissions limitations of reductions below the base year level, some have limitations at the base year level (i.e., no permitted increase above the base year level), while others have limitations above the base year level. Emission limits do not include emissions by international aviation and shipping. Although Belarus and Turkey are listed in the Convention's Annex I, they do not have emissions targets as they were not Annex I Parties when the Protocol was adopted. Kazakhstan does not have a target, but has declared that it wishes to become an Annex I Party to the Convention.

The 2010 Cancún agreements state that future global warming should be limited to below 2.0 °C relative to the pre-industrial level. During negotiations at various COPs to formulate a new arrangement after the expiry of the Kyoto Protocol, sharp differences in approach among various groups of countries reflecting their diverging interests became much more apparent. The developed countries were unwilling to continue to accept binding cuts on emissions due to the negative impact on their economies, and insisted that developing countries should also accept binding cuts in emissions. The latter group opposed such a major deviation from the basic principles of the UNFCC agreed at Rio, and insisted on the principle of common but differentiated responsibilities. Some analysts point out that the cumulative emissions of GHGs of the developed and developing countries have now almost evened out by 2010[21] and that it is time now for both groups to reduce emission. The other major issue was the question of finance and technology support to developing countries to enable them to reduce greenhouse gas emissions per unit of GDP while pursuing economic development. Ultimately, the Kyoto Protocol was extended in 2012 to cover the period 2013-2020 under the Doha Amendment adopted in December 2012, which still has not entered into force as of December 2015.

The Doha amendment contains binding targets on 37 countries. But Australia, the European Union (and its 28 member states), Belarus, Iceland, Kazakhstan, Liechtenstein, Norway, Switzerland, and Ukraine. Belarus, Kazakhstan and Ukraine have stated that they may withdraw from the Protocol or not put into legal force the Amendment with second round targets. Japan, New Zealand and Russia have participated in Kyoto's first-round but have not taken on new targets in the second commitment period. Other developed countries without second-round targets are Canada (which withdrew from the Kyoto Protocol in 2012) and the United States (which has not ratified the Protocol). As of July 2016, 66 states have accepted the Doha Amendment, while entry into force requires the acceptances of 144 states. Of the 37 countries with binding commitments, 7 have ratified. Therefore the Doha amendment has not been a success, and binding emission reduction targets have not been accepted by most of the developed countries.

## The Paris Agreement, 2015

Efforts have continued under various COPs to make progress, under pressure from concerned citizens groups over the impending prospects of large scale climatic disasters resulting from global warming. Ultimately the COP held in Paris in December 2015 succeeded in reaching agreement. This agreement seeks more ambitious actions to mitigate global warming and sets a target of 1.5 degrees Celsius average global temperature rise as the limit for global warming.

The agreement seeks to enhance the implementation of the UNFCCC through (a) Holding the increase in the global average temperature to well below 2 °C above pre-industrial levels and to pursue efforts to limit the temperature increase to 1.5 °C above pre-industrial levels, recognizing that this would significantly reduce the risks and impacts of climate change;(b) Increase the ability to adapt to the adverse impacts of climate change and foster climate resilience and low greenhouse gas emissions development, in a manner that does not threaten food production; and (c) Making finance flows consistent with a pathway towards low greenhouse gas emissions and climate-resilient development. Countries furthermore aim to reach "global peaking of greenhouse gas emissions as soon as possible".

The contribution that each individual country should make in order to achieve the worldwide goal are determined by all countries individually and called "nationally determined contributions" (NDCs). Article 3 requires them to be "ambitious", "represent a progression over time" and set "with the view to achieving the purpose of this Agreement". The contributions should be reported every five years and are to be registered by the UNFCCC Secretariat. Each further NDC should be more ambitious than the previous one, known as the principle of 'progression'. Countries can cooperate and pool their NDCs. The Intended Nationally Determined Contributions (INDC) pledged during the 2015 Climate Change Conference serve—unless provided otherwise—as the initial NDC of the Paris agreement.

The Paris agreement suffers from several weaknesses. The level of NDCs set by each country are not legally binding, as they lack the specificity, normative character, or obligatory language necessary to create binding norms. There is no mechanism to force a country to set a target in their NDC by a specific date and no enforcement if a set target in an NDC is not met. There will be only a "name and encourage" plan without any teeth. A trickle of nations exiting the agreement may trigger the withdrawal of more governments, bringing about a total collapse of the agreement. There are strong indications that the new US administration led by Donald Trump would after taking office in January 2017, move to reverse the actions taken by the previous US administration led by President Obama, on climate change, including the Paris agreement. The Paris Agreement has no provisions specifically on the question of providing finance and technology to de eloping countries to tackle climate change, a glaring omission.

The negotiators of the Paris Agreement stated that the NDCs and the 2 °C reduction target were insufficient to combat climate change, and proposed instead, a 1.5 °C target, noting "with concern that the estimated aggregate greenhouse gas emission levels in 2025 and 2030 resulting from the pledged INDCs do not fall within least-cost 2 °C scenarios but rather lead to a projected level of 55 gigatonnes in 2030", and recognizing furthermore "that much greater emission reduction efforts will be required in order to hold the increase in the global average temperature to below 2 °C by reducing emissions to 40 gigatonnes or to 1.5 °C". In other words, there is gap of 15 gigatonnes between what the INDCs add up to and what is required to meet the 1.5 degree Celsius target. The question is how is this gap to be met ?

Although not the sustained temperatures over the long term to which the Agreement refers, in the first half of 2016 average temperatures were about 1.3 °C above the average in 1880, when global record-keeping began.

## Mitigation, and the Carbon "space"

If there is to be a global limit on GHG emissions, then this must be shared among various countries. The developing countries still have much lower emissions per capita than the developed countries. Moreover, the latter have been adding to global emissions inventory for a much longer period, since the industrial revolution. Thus climate justice would imply that those countries that have added the most to global emissions should do the most now to reduce emissions. This was the basis of the principle of common but differentiated responsibility. Not only has this principle been diluted in recent years, but the developed countries have continued to press large developing countries such as India and China to cut emissions, even as these countries try and develop their economies which are still far below the developed world in terms of indicators such as per capita GDP.

To get a fair idea , it is necessary to examine data[22] on cumulative GHG emissions, recent annual GHG, GHG emissions per capita , and GHG emissions per unit of GDP PPP produced by the economy of some of the leading nations.

Table 11.3

**Data on GHG emissions cumulative, annual and per capita, and CO2 intensity of GDP**

| Country | Cumulative GHG emissions, billion tonnes, 1850-2010[23] | Recent annual GHG emissions[24], Million tonnes CO2 equivalent, 2012 | Per capita GHG emissions, Tonnes CO2 equivalent, 2012 | Kg CO2 emissions per $ PPP of GDP, 2012[25] |
|---|---|---|---|---|
| USA | 483 | 6235 | 20.67 | 0.317 |
| EU-28 | 443 | 4399 | 8.70 | 0.186 |
| Russia | 186 | 2322 | 16.24 | 0.506 |
| Japan | 73 | 1345 | 10.53 | 0.270 |
| Australia, NZ | 43 | 725 | 23.97 | 0.388 |
| Canada | 50 | 711 | 20.03 | 0.329 |
| Ukraine | 39 | 390 | 8.55 | 0.765 |
| China | 299 | 10976 | 8.13 | 0.654 |
| India | 106 | 3014 | 2.39 | 0.325 |
| Brazil | 101 | 1013 | 5.22 | 0.152 |
| World total | 2585 | 44816 | 6.35 | 0.356 |

Most countries showed a significant improvement in $CO_2$ intensity of GDP, as a result of measures taken to limit emissions and improve energy efficiency. The best performance was shown by the EU-28 group, improving intensity from 0.57 in 1990 to 0.19 in 2013, while the world average improved from 0.78 to 0.34 in the same period. The EU-28 figure sets a benchmark for the results that could be obtained with optimum use of technology, finance and policy measures. Breakthroughs in technology would result in a further improvement in intensity. For example, France which relies heavily on nuclear power has achieved an intensity of 0.129 in 2013, while Germany which has a highly developed renewable energy sector has achieved 0.208. This shows the vast scope for improvement in the development path of developing countries including India and China, so that GDP growth can take place while limiting emissions. However this is predicated on technology and finance being available.

## Alternative Energy Sources and Technology

The above illustrates the crucial importance of technology for alternative energy, energy efficiency, and carbon reduction in achieving the goal of limiting global warming. Technology breakthroughs appear to be within reach in solar power and energy efficiency especially in lighting. Technology for removal of $CO_2$ from the atmosphere (Carbon capture and sequestration (CCS) is an area that deserves far more support than has been provided so far, as it can enlarge the carbon space available for development. It is of course axiomatic that whatever technology emerges should be easily accessible to the developing countries on reasonable terms.

## Financing Mechanisms

The other crucial plank for fighting global warming is the question of financing for climate change mitigation and adaptation efforts. The global climate finance architecture is complex: finance is channeled through multilateral funds – such as the Global Environment Facility and the Climate Investment Funds – as well as increasingly through bilateral channels. In addition, a growing number of recipient countries have set up national climate change funds that receive funding from multiple developed countries in an effort to coordinate and align donor interests with national priorities.

Climate finance is being provided by national, regional and international entities for support mechanisms and financial aid for mitigation and adaptation activities to promote the transition towards low-carbon, climate-resilient growth and development through capacity building, R&D and economic development. The term has been originally used to refer to transfers of public resources from developed to developing countries, in light of their UN Climate Convention obligations to provide "new and additional financial resources," but has now been widened to refer to all financial flows relating to climate change mitigation and adaptation.

Finance may be provided by public, private and public-private sectors and can be channeled through various intermediaries, notably bilateral or multilateral financial and development cooperation agencies, the UNFCCC (funds such as those managed by the Global Environment Facility), non-governmental organizations and the private sector. The financial flows can flow from developed to developing countries (North-South), from developing to developing countries (South-South), from developed to developed countries (North-North) and domestic climate finance flows in developed and developing countries. According to some reports, investments in renewable energy in 2010 reached a record of USD 211 billion (not including large hydropower). The preliminary estimates of financing needs for mitigation and adaptation activities in developing countries range from USD 140-175 billion per year for mitigation over the next 20 years with associated financing needs of USD 265-565 billion and USD30 – 100 billion a year over the period 2010 - 2050 for adaptation.

A number of initiatives are underway to monitor and track flows of international climate finance. The UNFCC carries out an overview of Climate Finance Flows. There is the need to make the monitoring of these flows more effective, such as harmonizing and consistent reporting of data, and transparency over implementation of projects and programmes over time. A large number of initiatives that have been implemented to assist developing countries to manage their response to climate change, both through information provision and policy-relevant analysis. More transparency is needed about the status of implementation bilateral climate finance initiatives. The proliferation of climate finance mechanisms increases the challenges of coordinating and accessing finance.

The Global Environment Facility (GEF) launched by the World Bank in 1991, involves 183 member countries in partnership with international institutions, civil society organizations (CSOs), and the private sector to address global environmental issues while supporting national sustainable development initiatives. Today the GEF is the largest public funder of projects to improve the global environment and provides grants for projects related to biodiversity, climate change, international waters, land degradation, the ozone layer, and persistent organic pollutants. Since 1991, the GEF has achieved a strong track record with developing countries and countries with economies in transition, providing $12.5 billion in grants and leveraging $58 billion in co-financing for over 3,690 projects in over 165 countries. However the GEF's resources are far too small and dispersed over too many activities to meet the needs of climate finance.

The Copenhagen Accord, established at COP-15 in Copenhagen in 2009 mentioned the "Copenhagen Green Climate Fund". The Fund was formally established during the 2010 COP-16 at Cancun and is based in Songdo, South Korea, within the UNFCCC framework. Its governing instrument was adopted at the 2011 COP 17 in Durban, South Africa. COP 17 decided that the "GCF would become an operating entity of the financial mechanism" of the UNFCCC, and that on COP-18 in 2012, the necessary rules should be adopted to ensure that the GCF "is accountable to and functions under the guidance of the COP". The Green Climate Fund was intended to be main instrument of Long Term Financing under the UNFCCC, which has set itself a goal of raising $100 billion per year by 2020. Uncertainty over where this money would come from, the lack of pledged funds and potential reliance role of the private sector are unsettled issues. Other issues are the balance between support to adaptation and mitigation. In view of these problems the GCF has not been a success so far.

## The Emissions Gap

As mentioned there is a shortfall in the total of INDCs declared by countries and the emissions reductions that are required to meet the target of keeping global warming. This "emissions gap" has been the focus of reporting by UNEP, since meeting this gap is critical to combating global warming. Global GHG emissions by 2030 based on the latest UNEP assessment of the INDCs submitted by countries is in the range of 51-57 $GtCO_2e$ compared to 58-62 $GtCO_2e$ if current policies continue. [26]. But the global emission levels in 2030 required for a likely chance (>66 per cent) of staying below the 2°C goal in 2100, following a least-cost pathway from 2020 with only modest improvement of the GHG intensity until then, is 31-44 $GtCO_2e$. If the warming is to be kept within 1.5 degrees Celsius, then the emission level needs to be further reduced to 39 $GtCO_2e$ by 2030.

The present level of GHG emissions in 2014, the latest available figure was 53 $GtCO_2e$ with an annual growth rate of 2.2 percent during 2000-2010. In 2015 global $CO_2$ emissions stagnated for the first time and showed signs of a weak decline compared to 2014 (-0.1 per cent). This was preceded by a slowdown in the growth rate of $CO_2$ emissions, from 2.0 per cent in 2013 to 1.1 per cent in 2014. However this may also be the impact of slowdown in global economic growth apart from environment related actions.

Thus there is a gap between what is required and the current trends of emissions of some 17 $GtCO_2e$ by 2030, which will need to be bridged. The IPCC in its fifth report had concluded that to meet the 2 degree Celsius target, the remaining cumulative CO2 emissions, the so called carbon budget for 2011 onwards, must be kept within 1000 GtCO2e. The implications of all these estimates is that the target of 2 degrees Celsius requires global GHG emission to be reduced to zero by 2016-2075, which will require concerted and sustained all out action. It may also be noted that the INDCs are only voluntary pledges and statements of intentions, and there is no built in mechanism to ensure that they are met. The chances for effective action by the international community seem remote, unless some large scale disasters with severe economic consequences drive it to take action. In the present scenario, there seems to be a lack of political will among countries to take the required actions, and insufficient understanding of the emissions gap and its consequences. More public information is required to explain the situation in simple terms.

## Low Emission Economy, Taxes and Incentives

Obviously the solution to global warming can come only through a transition to a lower emission economy. This will require the complete reengineering of all processes to reduce emissions through use of new technology. From energy production, distribution, consumption, transportation, manufacturing, agriculture and food, and services, all sectors will need action. This will be an extremely complex task and may not be possible without economic and market based incentives. One of the incentives could be a system of Carbon taxes[27], and credits for Carbon reductions. The latter has already been in existence since the Kyoto Protocol. Carbon tax appears to be a logical way forward, but political will to introduce such a tax is lacking. The revenues from the Carbon tax could be used to provide incentives for carbon reduction and lower carbon technology introduction. Carbon related import tariffs could also be levied for imported products based on carbon emissions, so that lower carbon production is not unfairly disadvantaged. Over the long run flexible and sensible market based incentives and taxes could bring about a transition to a lower emission economy. It would be a wider application of the long accepted principle that the polluter pays.

## India's Role

India has a major role to play in limiting global warming. Its sheer size and population and its rapid economic growth needs to be accommodated into any global framework. At this stage India needs a rapidly growing economy to lift its population out of poverty and assure a decent standard of living. Should India follow the same type of economic growth that the developed countries have done in the past, its per capita carbon emissions and total GHG emission would skyrocket to a level that would negate all efforts to limit global warming. Besides, India does not have the required level of fossil energy resources to meet its needs, and cannot sustainably import these resources. Therefore India has to follow a more sustainable, lower emissions pathway for its development. This is reflected in national policy statements and measures and in India's submission on INDCs[28]. India has also launched a number of initiatives for renewable energy, such as solar energy, and civil nuclear energy expansion.

India declared a voluntary goal of reducing the emissions intensity of its GDP by 20–25%, over 2005 levels, by 2020. Emission intensity of India's GDP has decreased by 12% between 2005 and 2010. India's INDC submission includes commitments such as (1) reducing the emissions intensity of its GDP by 33 to 35 percent by 2030 from 2005 level; (2) 40 percent cumulative electric power installed capacity from non-fossil fuel based energy resources by 2030 and (3) create an additional carbon sink of 2.5 to 3 billion tonnes of $CO_2$ equivalent through additional forest and tree cover by 2030. Given access to technology and finance and international support, India would certainly do its part to meet the challenge of global warming.

## Solar Energy in India

India is amply endowed with solar energy, due to its location near the equator. Solar energy is therefore a huge resource that can meet India's growing energy needs and give it energy security. In recent years India has made considerable progress in harnessing solar energy. The Indian solar sector scored impressive growth[29] with cumulative installations reaching approximately 12.8 GW at the end of 2017, as against the revised target of 100 GW by 2022. India has become one of the top solar markets in the world after China and the United States. Solar tariffs have declined by about 75 percent since 2010, reaching a new record low of Rs.3.15 (4.8 US cents)/kWh. Rooftop installations in India have totaled nearly 850 MW as of 2017. The government is targeting 40 GW by 2022 through rooftop installations, but policy support currently is minimal. Over 20 states have some net-metering policy, but few have a functioning net-metering program. Other issues affecting rooftop installations include lower accelerated depreciation, and the removal of the 10-year income tax holiday. Nevertheless, the future prospects for solar power in India look bright, driven by high demand for power, and further reductions in capital costs due to technological development.

## International Solar Alliance Initiative

India has taken a more active role on climate change issues. At the Paris Climate Summit in November 2015, Prime Minister Modi along with French President Francois Hollande, announced a major initiative, the International Solar Alliance (ISA)[30]. This marks a new effort by India to play a more proactive and constructive role in fighting climate change, and embrace renewable energy, including solar power, in a big way, and to use foreign policy to support its leadership role among developing countries. India sees solar energy as a major source of clean energy for India that can meet its huge requirements for energy. The ISA is an alliance of 121 countries situated between the Tropics of Cancer and Capricorn that enjoy 300 days of sunshine and have large solar energy potential. In January

2016, the two leaders laid the foundation stone of ISA at Gurugram near New Delhi. The World Bank signed an agreement with the ISA in 2016 to mobilize $1 trillion in investments by 2030. ISA is part of India plans to generate energy through renewables. The target for solar power generation has been increased from 20 GW to 100 GW by 2022. ISA could also help India to develop technologies and equipment for its domestic market as well as sell to countries with similar needs.

The ISA framework agreement was opened for signing up at the Conference of the Parties at Marrakesh in November 2016 and 25 countries including France, Bangladesh, Brazil and Tanzania have joined it. The assembly will meet after 15 of these signatories ratify ISA. The alliance will have an assembly, a council and a secretariat. The Indian government will support the secretariat for five years, after which it would have to generate its own resources. The secretariat has been set up at the National Institute of Solar Energy in Gurugram. The ISA will also collaborate with other multilateral bodies such as the International Energy Agency, the International Renewable Energy Agency and the United Nations. The ISA would be the first such treaty-based international government organization headquartered in India.

## Future Outlook and Prospects

There is much more scientific information and widespread public awareness today about the threats posed by global warming and climate change. This will increase as extreme climate events continue to unfold across the globe. However, political will to address this challenge is still insufficient both at the national as well as international level. The emissions gap reports reveal starkly the gap which needs to be filled by concrete actions. The investment in research and technology specifically to deal with global warming needs to be stepped up manifold[31]. Market based incentives for emissions reduction and taxes for emission should be widely introduced and applied. Monitoring of progress towards a lower emissions economy should be intensified. For the developing countries, the choice of a lower emissions intensity pathway to development must be promoted and encouraged. Countries and regions suffering from climate change must be helped to overcome these challenges. National and international policy measures are still evolving and need further and urgent development. While the Paris agreement of 2016 offers hope it remains to be seen whether its promise will be fulfilled. There is much more to be done by way of international cooperation in this complex and multidisciplinary field. Climate change has become the major challenge of the 21st century. The UN has identified it as Goal 13 of the 17 Sustainable Development Goals for 2030, to "take urgent action to combat climate change and its impacts."

## Endnotes

1   The sun's evolution, David Taylor, 2012 http://faculty.wcas. northwestern.edu/~infocom/The%20Website/evolution.html , accessed 26-10-2016

2   Composition of the Earth's atmosphere today, Universe Today, Fraser Cain, 2015 http://www.universetoday.com/26656/composition-of-the-earths-atmosphere , accessed 26-10-2016

3   Earth's annual global mean energy budget. Kiehl J T et al, Bulletin of the American Meteorological Society. 78 (2): 197–208, 1997, https://web.archive.org/web/20060330013311/http://www.atmo.arizona.edu/students/courselinks/spring04/atmo451b/pdf/RadiationBudget.pdf , accessed 26-10-2016

4   Methane emissions and climatic warming risk from hydraulic fracturing and shale gas development: implications for policy, Howarth R. W., 2015, https://www.dovepress.com/methane-emissions-and-climatic-warming-risk-from-hydraulic-fracturing--peer-reviewed-article-EECT , accessed 26-10-2016

5   In fact ozone depletion may actually produce a small negative change in the greenhouse effect.

6   Dobson Unit (DU) is a measure of total-column ozone, expressed as the thickness (in units of 10 μm) of that layer which would be formed by the total gas in a column under standard temperature and pressure. 1 DU is approximately 1.25 ppb of the average total column of air.

7   The evolution of policy responses to stratospheric ozone depletion. Morrisette, P. M., Natural Resources Journal 29: 793-820, 1989 http://www.ciesin.org/docs/003-006/003-006.html , accessed 9-11-2016

8   The Toronto Group sought controls that would force the European countries to cut back on aerosol use of CFCs, while the EEC opposed being forced to adopt regulations already adopted by the Toronto Group countries. The dispute polarized the negotiations.

9   This concept would find a key place in other environmental negotiations, such as over climate change. The developed countries have passed through the period of industrialisation during which they have released a larger total quantity of damaging substances, while the developing countries still have to industrialise and need greater time to adapt. It also happens that ozone layer depletion had a greater impact on developed countries.

10  Multilateral Fund for the Implementation of the Montreal Protocol, http://www.multilateralfund.org , accessed 11-11-2016. This Fund has provided over $3.3 billion of assistance since it was set up in 1991.

11  It has been revised eight times - in 1990 (London), 1991 (Nairobi), 1992 (Copenhagen), 1993 (Bangkok), 1995 (Vienna), 1997 (Montreal), 1998 (Australia), 1999 (Beijing) and 2016 (Kigali, adopted, but not in force). http://ozone.unep.org/sites/ozone/files/pdfs/Consolidated-Montreal-Protocol-November-2016.pdf . accessed 11-11-2016

12  197 countries have ratified the Protocol, practically the entire UN membership. The ozone layer has been recovering steadily since the Protocol was implemented and is expected to recover completely by

around 2065, according to experts assessment. Environmental effects of ozone depletion and its interactions with climate change:2014 assessment, http://ozone.unep.org/Assessment_Panels/EEAP/eeap_report_2014.pdf , accessed 11-11-2016

13 Recent Greenhouse Gas Concentrations, Blasing, T.J. , April 2016, http://cdiac.ornl.gov/pns/current_ghg.html , accessed 14-11-2016

14 The calculations of radiative forcing and its actual impact on global temperature involve extremely complex global climate models and the actual global temperature rise from these models varies significantly. Further refinement of global climate models should reduce these variations in future.

15 Permafrost melting and release of methane becomes increasingly likely as the global temperature rises by 1.8 degrees Celsius. For this reason some experts advocate action to keep global temperature rise below 1.5 degrees Celsius.

16 Venting and leaking of methane from shale gas development, R. W. Howarth et al, Climate Change, 2012, http://www.eeb.cornell.edu/howarth/publications/Howarthetal2012_Final.pdf , accessed 14-11-2016

17 Fracking Would Emit Large Quantities of Greenhouse Gases, Mark Fischetti, Scientific American, 2012 https://www.scientificamerican.com/article/fracking-would-emit-methane , accessed 14-11-2016

18 A large increase in U.S. methane emissions over the past decade inferred from satellite data and surface observations, Turner, A. J. et al, Geophysical Research Letters, April 2016, http://onlinelibrary.wiley.com/doi/10.1002/2016GL067987/epdf , accessed 14-11-2016

19 Climate Change 2014 Synthesis Report, Summary for Policymakers, IPCC, 2014 https://www.ipcc.ch/pdf/assessment-report/ar5/syr/AR5_SYR_FINAL_SPM.pdf , accessed 14-11-2014

20 Where there are threats of serious or irreversible damage, lack of full scientific certainty shall not be used as a reason for postponing cost-

effective measures to prevent environmental degradation. This is embodied in the Rio , principle 15

21 During 1850-2010 the share of cumulative emissions of GHGs from developed and developing countries was estimated at 52 percent and 48 percent respectively. See 22 below for details.

22 Countries' contributions to climate change, Michel G. J. den Elzen et al, Climatic Change, 31 October 2013, http://www.pbl.nl/en/ publications/countries-contributions-to-climate-change , accessed 30-12-2016

23 This figure includes all GHGs, converted into CO2 equivalent. The period chosen is 1850-2010. There is a rapid increase in GHGs during the period 1950-2010 from the developing countries such as China and India, reflecting their rapid economic growth.

24 World Research Institute, CAIT Tools. Figures exclude land use and forestry contributions.

25 World Bank, World Data Bank, CO2 emissions (kg per PPP $ of GDP). This is the CO2 intensity of GDP PPP which indicates the efficiency of the economy in using energy. Almost all countries have shown significant improvements in this parameter since 1990, according to data. India improved from 0.621 in 1990 to 0.302 in 2013, almost doubling its efficiency, while China improved from 2.20 in 1990 to 0,61 in 2013.

26 The Emissions Gap Report 2016 - A UNEP Synthesis Report , UNEP; https://uneplive.unep.org/media/docs/theme/13/Emissions_Gap_ Report_2016.pdf , accessed 3-1-2017

27 A number of countries have already introduced such a tax. British Columbia, Canada introduced a tax of C$10 per tonne of CO2 in 2008, increasing up to C$30 per tonne by 2012. There was a 12.9% decrease in British Columbia's per capita emissions in 2008-2013 compared to 2000-2007 some three-and-a-half times as pronounced as the 3.7% per capita decline for the rest of Canada. See Where Carbon

is taxed, https://www.carbontax.org/where-carbon-is-taxed/ , accessed 3-1-2017

28 India's intended nationally determined contribution: Working towards Climate justice, October 2015 http://www4.unfccc.int/submissions/ INDC/Published%20Documents/India/1/INDIA%20INDC%20 TO%20UNFCCC.pdf , accessed 3-1-2017

29 Solar sector seen adding 10 GW this year, Mercom Capital, 28 Apr 2017, http://mercomcapital.com/a-record-year-for-india-forecasted-with-approximately-10-gw-of-solar-installations-in-2017, accessed 28-4-20-17

30 With International Solar Alliance, India seeks its place under the sun, E. Roche, Livemint.com, 28 Apr 2017, http://www.livemint.com/ Industry/UVkC61xw2x6WGoWQSGjkFN/With-International-Solar-Alliance-India-seeks-its-place-und.html , accessed 28-4-2017

31 Tenfold jump in green tech needed to meet global emissions targets, Duke University, Science Daily, 3 January 2017, https://www. sciencedaily.com/releases/2017/01/170103152452.htm, accessed 5-1-2017

# Chapter 12

# Human Health – a Healthy World for all

*"It is more important to know what sort of person has a disease than to know what sort of disease a person has"*

*– Hippocrates*

## Introduction

Advances in science and technology have had a major impact on human health. This has also resulted in health issues gaining prominence in international relations. Health problems know no national boundaries, and advances in medicine are of interest to all nations. In the health sector, there has been a constant conflict between commercial interests and public health, accentuated by new technological developments. This has been an important influence on international cooperation in this sector.

International cooperation in health has a long history, starting from the International Sanitary Conference in Paris in 1851, largely to control the spread of diseases such as cholera across national boundaries through quarantine measures. After the Second World War, it was decided to create the World Health Organization which came formally into existence[1] in 1948 with 55 member states and headquarters in Geneva. Its first priorities were to control the spread of malaria, tuberculosis and sexually transmitted infections, and to improve maternal and child health, nutrition and environmental hygiene. The WHO currently has 194 members as on 2015, and its proposed budget has increased over the years reaching about US$4 billion in 2014-15. About US$930 million of this is provided by member states with a further US$3 billion from voluntary contributions.

While the WHO is the central coordinating agency of the UN system in the health sector, a number of other international bodies are also active in this sector. These include the World Bank, the UNICEF, UNDP, etc. A large number of bilateral and non-governmental organizations are also active in this sector. There a growing number of issues where health cooperation at the international level has become increasingly important over the years. These are briefly covered below.

## Narcotics and Psychotropic Drugs Control

Narcotics and psychoactive substances have for long been known to humans. These natural products such as coca, marijuana, opium have been in used and abused for centuries and have been the source of economic benefits to those involved in production, distribution and consumption of these substances. The colonial countries especially such as the UK and Netherlands derived great benefits from narcotics trading activities. Indeed some authors have described the British opium trade between India and China[2] as nothing more than a large scale organized pushing of opium produced in India into China for large profits.

The US initiated efforts at prohibition based international drug control as the effects of drug addiction and abuse grew at home, and pressure from religious groups, while the colonial powers resisted such efforts but ultimately were forced to go along. The efforts focused on prohibition and attempted to stem the flow of drugs into their territories. In doing so, they earned political capital back home and shifted the cost and burden of drug control to predominantly Asian and Latin American developing countries. This also stimulated the growth and development of the global illicit drug trade. There was little success in controlling the supply of drugs at the source.

Gradually a prohibition based international drug control system came into existence starting from 1909. Advances in chemical technology led to a proliferation of narcotics substances derived from natural products as well as synthetic products. This required constant updating of control arrangements.

The current legal and administrative framework for international drug control[3] is laid out in three international Conventions negotiated under the auspices of the United Nations (UN): (1) the Single Convention on Narcotic Drugs, 1961 (Single Convention) as amended by the Protocol Amending the Single Convention on Narcotic Drugs,1961; (2) the Convention on Psychotropic Substances (Psychotropics Convention); (3) The Convention against Illicit Traffic in Narcotic Drugs and Psychotropic Substances (Trafficking Convention).

The Single Convention is a consolidation of nine multilateral drug control treaties negotiated between 1912 and 1953. The Convention extended the control system to include the raw materials for narcotics. Well over 100 narcotic drugs are controlled under the Convention. These include primarily plant based products such as opium, opium derivatives (morphine, heroin, and codeine), cannabis, coca and cocaine, but also synthetic narcotics including methadone and pethidine. The substances are categorized into four schedules, each schedule being subject to a different level of control. The Single Convention prohibits, for example, opium smoking and eating, coca leaf chewing, cannabis resin smoking, and the nonmedical use of cannabis. The Single Convention Protocol further tightens controls on the production, use and distribution of illicit narcotics. The Protocol also contains provisions on treatment and rehabilitation for drug abuse and addiction.

The Psychotropics Convention extends international control to include numerous synthetic psychotropic substances: stimulants, such as amphetamines, depressants, including barbiturates, and hallucinogens, such as mescaline and LSD (lysergic acid diethylamide). Similar to the Single Convention, the drugs are organized into four schedules depending on their addiction and abuse potential, and their therapeutic value. The Convention sets out detailed provisions concerning the international trade of psychotropics, including measures that strictly control their export and import. Measures for the prevention of abuse and for rehabilitation are also included.

The Psychotropics Convention imposes significantly weaker controls due to pressures from the multinational pharmaceutical industry. The group of countries with large pharmaceutical capacity advocated weak controls, and national sovereignty taking precedence over a strong supranational UN body. The Psychotropics Convention makes no mention of the "serious evil" of "addiction," but rather notes "with concern the public health and social problems resulting from the abuse of certain psychotropic substances." As well, it is recognized that "the use of psychotropic substances for medical and scientific purposes is indispensable and that their availability for such purposes should not be unduly restricted." The general tone of the Psychotropics Convention Preamble is less harsh, and it is implied that "abuse of certain," not all, psychotropics, is not as serious a problem as "addiction to narcotic drugs" in general.

The Trafficking Convention seeks to combat the illicit traffic of drugs. Its key goals are improved international law enforcement cooperation and strengthened domestic criminal legislation. The Convention contains provisions on money laundering, the freezing of financial and commercial records, extradition of drug traffickers, and transfer of criminal proceedings, mutual legal assistance, and strict monitoring of chemicals often used in illicit production.

India is a signatory to the single Convention on Narcotic Drugs 1961, as amended by the 1972 Protocol, the Conventions on Psychotropic Substances, 1971 and the United Nations Convention against Illicit Traffic in Narcotic Drugs and Psychotropic Substances, 1988. Under the Narcotic Drugs and Psychotropic Substances Act of 1985, the Narcotics Control Bureau was set up in 1986 as the principal coordinating agency for drug control in India. India is one of the largest legal (licit) producers of opium poppy and the only country which legally produces opium gum. Production is regulated through a system of annual licences and is sold at designated prices to a government agency. Licit opium production is intended for use by the drug industry for processing into various products for pain relief.

The prohibition based strategy of drug control has not succeeded in curbing the illegal drug trade which continues to flourish, especially in areas where there is weak or non-effective government in control, such as conflict zones. In addition the harsh penal measures introduced in many countries have provoked opposition from human rights groups who argue that the penalties are disproportionate to the crime, especially in the case of small or casual users of drugs. Thus there is a need to balance control efforts between prohibition and demand reduction.

The demand pressure for psychotropic drugs has led to technology being used to circumvent controls and to develop synthetic drugs such as Ecstasy, LSD and methamphetamine. In addition there are Spice or K2 (synthetic marijuana), Bath Salts (synthetic stimulants) and a drug known as "N-bomb"[4], known as "designer drugs." A designer drug is a synthetic (chemically made) version of an illegal drug that was slightly altered to avoid having it classified as illegal. As law enforcement catches up with new chemicals that are so created and makes them illegal, manufacturers devise altered versions to steer clear of the law.

Some of these drugs are sold over the Internet or in certain stores (as "herbal smoking blends"), while others are disguised as products labeled "not for human consumption" (such as "herbal incense," "plant food," "bath salts" or "jewelry cleaner") to mask their intended purpose and avoid health and safety rules. Due to the constantly growing number of chemicals that are developed, designer drug users have no way of knowing what the drugs they take might contain. Further, as a small modification made to a known drug may result in a new drug with greatly different effects, and unpredictable risks to health of users.

India is a major player in this field. It is a major licit producer of opium. Its population is large and has over millennia been used to natural plants with psychotropic effects. It has several countries in its neighbourhood where illicit production of drugs is flourishing, leading to a risk of becoming a transit or consuming country. Some areas of India already have large drug abuse problems, such as Punjab and the North East. Synthetic drug abuse[5] is also becoming a serious problem given India's capacity for drug manufacturing, and relatively weak controls. The nexus between organized crime, drug trafficking and terrorism is also a security risk. Therefore this is a sector where India will have to be actively engaged in future, cutting across many government agencies.

**Access to Essential Drugs and Medicines**

The conflict between commercial interests and public health is sharply evident in the field of access to essential drugs and medicines at affordable prices. The TRIPs agreement of 1995 effectively recognized product patents[6] and blocked the avenue of reverse engineering used in countries such as India to find alternative, cheaper ways of producing patented drugs. It also increased the scope of patent protection and reduced the possibilities for compulsory licensing in the public interest, although the Doha Declaration of 2001 restored some balance. Countries such as India were forced to amend their patent laws to conform to TRIPs provisions. This led to a rise in prices of most patented drugs. The Doha Declaration came about partly as a result of public pressure in US and Europe which finally resulted in drug companies dropping in 2001 the case against South Africa over its legislation to increase the availability of affordable medicines.

Another example is of Brazil's AIDS programme which was successful due to its ability to produce medicines locally. Brazil had also been able to negotiate lower prices for patented drugs by using the threat of production under a compulsory license. Brazil had offered a cooperation agreement, including technology transfer, to developing countries for the production of generic ARV drugs. In February 2001, the United States took action against Brazil at the WTO Dispute Settlement Body. The United States action came under fierce criticism from the international NGO community, and ultimately in June 2001, the US withdrew the complaint.

The "essential drugs concept (EDC)" is the relatively few pharmaceuticals necessary for the preservation of health. Expert panels convened by WHO have determined that only about 200–250 drugs and vaccines are truly necessary to relieve most health problems. The cost of drugs in developed and less-developed countries was and is high, and many drugs were either useless or harmful. There are also "irrational combinations", drugs that separately are beneficial but have no use in combination. Many pharmaceutical manufacturers felt threatened by the EDC, which called into question some of their most lucrative preparations. Doctors, often educated about drugs or lobbied by drug manufacturers, resented the imagined interference with their autonomy. Both groups have opposed the adoption of the EDC. Drug manufacturers used their influence to threaten developing countries that adopt the EDC with economic or political sanctions. The EDC concept is part of a broader concept of rational use of drugs, which focuses on the most appropriate and effective use of drugs to combat illness. The EDC and WHOs model lists of essential drugs and medicines issued periodically has enabled governments in developing countries to substantially reduce the inventory costs of stocking drugs in their public health systems and weed out non essential drugs. The focus on EDC has also enabled reductions in prices and helped bring about access to these drugs and affordable prices.

As basic health standards improve and life spans increase, the incidence of non communicable diseases such as cardiovascular disease, neurological diseases, and cancer will increase even in developing countries. The cost of treatment for cancer is extremely high, and the available drugs are beyond the reach of many people. The cost of research and development of anti cancer drugs is also extremely high. This raises the questions of how to make anti cancer drugs accessible to all. The present system, which relies largely on insurance and the markets, has many deficiencies and gaps. Greater public financing and direct state institutional support for research and development of anti cancer drugs is clearly necessary. Technology breakthroughs in understanding of cancer leading to development of new strategies to fight this disease should lead to reduction in costs of cancer treatment in the future.

## Health care for all

Health for All (HFA) has been the subject of international debate since the WHO sponsored Alma Ata declaration of 1978, which set the target of achieving health for all by 2000, a goal that is still far from being reached. The key concept was that all people would have access to primary health care. The Alma-Ata Declaration called for education about hygiene, health, and prevention; education about nutrition, food supply, and water supply; child and maternal health care, including family planning; immunization against major diseases and prevention of endemic local maladies; treatment of disease appropriate to the local conditions; and the provision of essential drugs. Many countries made significant progress towards this goal. The main constraint was the lack of funding for public health efforts and weaknesses in capacity of the public health care system. Universal affordable health care and appropriate insurance coverage for all has become an important political issue in several advanced countries such as the USA. Technology breakthroughs may well make it easier to achieve this goal, provided the relevant technology is made available and is accessible to the countries and groups that need it. The benefits of medical research should not result in health care going beyond the reach of those who need it across the world.

## HIV and AIDS

During the 1980s AIDs emerged as an untreatable disease spreading rapidly across the world. Research led to the finding in 1983 that the disease was caused by a virus, the Human Immunodeficiency Virus (HIV) that probably originated in West African primates. Currently there is no treatment to cure or vaccine to prevent this disease, despite intensive efforts. However antiretroviral treatment is available which can slow the progression of the disease. Prevention was another important strategy relying on public education and counseling, which also helped to stop the spread of misinformation about AIDs.

International efforts to control AIDs grew rapidly as country after country reported growing number of cases. In 1996 a joint venture, UNAIDS based in Geneva, was launched with the partnership of 11 UN-system agencies to combat HIV/AIDS, truly a global pandemic. In 2015 it had a total budget of $ 224 million. The Global Fund to Fight AIDS, Tuberculosis and Malaria (or the Global Fund) was set up as an international financing organization that aims to mobilize resources to prevent and treat HIV and AIDS, tuberculosis and malaria. A public–private partnership, the Geneva based organization started operations in January 2002. Microsoft founder Bill Gates was one of the first private foundations among many bilateral donors to provide seed money for the project. As of July 2016, the organization had disbursed $30 billion to countries and communities in need. According to the organization, it is supporting 9.2 million people on antiretroviral therapy for AIDS.

As of 2014, approximately 37 million people have HIV worldwide[7] with the number of new infections in 2014 being about 2 million, down from a peak of 3.1 million in 2001. Global coverage of antiretroviral therapy has steadily increased, from 22 percent in 2010, reaching 46% at the end of 2015. The number of AIDS related deaths per year has decreased from a peak of 2 million in 2005 to about 1.2 million in 2015. The world has committed to the ambitious goal of ending the AIDS epidemic by 2030 as part of the UN's Sustainable Development Goals.

## New and Emerging Diseases

The recent outbreak of Ebola in 2013-2016, centred in West Africa has vividly brought out the threat of new and re-emerging diseases. The outbreak caused a total of 28,616 suspected cases and 11,310 deaths (39.5%), though the WHO believes that this substantially understates the magnitude of the outbreak. The outbreak exposed several weaknesses in the global systems for responding to such outbreaks and led to significant improvements. An expert group's recommendations included measures to prevent small outbreaks from turning into large outbreaks, and demanded encouragement of early international reporting of outbreaks by following agreed international rules and stronger operational capacity within the WHO as well as the aid system, if outbreaks turned into emergencies. Furthermore, mobilization of the understanding needed to fight outbreaks would require an international structure of rules to enable access to the benefits of research, and financing to establish technology when commercial motivations were not appropriate. The report also noted that competent governance of the global system demanded political leadership and reform of the WHO.

The WHO has identified the following emerging diseases needing urgent R&D attention - Crimean Congo haemorrhagic fever, Ebola virus disease and Marburg, Lassa fever, MERS and SARS coronavirus diseases, Nipah and Rift Valley fever. WHO's plan includes focusing accelerated R&D on dangerous pathogens which are the most prone to generate epidemics. As well as advocating for the initiation or enhancement of the R&D process to develop diagnostics, vaccines and therapeutics for these diseases, the plan will also consider behavioural interventions, and filling critical gaps in scientific knowledge to allow the design of better disease control measures.

By 2050, the world's population will have risen to 9.7 billion. Cities will become increasingly dense and shanty towns with inadequate housing and a lack of basic services such as water, sewerage and waste management will grow. A combination of high population density, poverty, changes in social structures, and a lack of public health infrastructure will create progressively more favourable conditions for emerging diseases. The increasing global flow of commodities, people and animals coupled with increased population density will magnify the transmission of diseases, both between people and across the human-animal barrier. Millions of passengers travel: over 2 billion global passengers travelled annually by air in the first decade of the 21st century, compared with just 68.5 million in the 1950s. Continued growth in the movement of people and commodities between urban centres intensifies the risk of infectious transmissions across geographies and diminishes the ability to respond to, and prepare for, a global disease outbreak. There are some 20 known infectious diseases that have re-emerged or spread geographically, including dengue, chikungunya, typhoid, West Nile, artemisinin-resistant malaria and the plague. Other known threats – such as influenza (i.e. H1N1 Swine Flu), MERS-Cov, and Ebola – continue to raise concerns. Even when known infectious diseases can be mitigated by existing treatments or vaccines, there is the risk of emerging resistant strains, mutating viruses, or a pandemic that is so large it renders response inadequate.

Faced with this challenge it makes eminent sense to develop international cooperation covering research and development, and disease control mechanisms and capacities, so that rapid and effective responses can be mounted against any such threat. It also implies that basic health infrastructure in all countries must be of a minimum level of effectiveness, since this is the system that will be at the forefront of fighting an outbreak. The World Bank is planning a Pandemic Emergency Financing Facility (PEF) to respond more rapidly to emergencies caused by outbreaks. WHO has set up the Global Outbreak Alert and Response Network (GOARN) to ensure that technical expertise and are skills are on the ground where and when they are needed most during disease outbreaks. GOARN is a collaboration of existing institutions and networks, constantly alert and ready to respond. The network pools human and technical resources for rapid identification, confirmation and response to outbreaks of international importance.

Surveillance of emerging infectious diseases is vital for the early identification of public health threats. New technologies are becoming available for the rapid genomic identification of pathogens and development of vaccines, but also for the more accurate monitoring of infectious disease activity including web-based surveillance tools. Nevertheless the challenge of combating new and re-emerging diseases will continue to demand resources and effort by the international community in the future.

## Chemical Safety and Security

The accident at the Union Carbide chemical plant in Bhopal, India on 2-3 December 1984 will remain etched in memory as one of the worst ever chemical disasters. The leakage of some 40 tonnes of poisonous methyl isocyanate gas from a storage tank affected over 500,000 people living nearby turning parts of Bhopal into a virtual gas chamber, and caused deaths of 3800 people and severely affected another 42,000, according to official reports. There has been much discussion over the issue of liability and compensation to the injured and killed persons, and the medical impact of the exposure to the gas. However, the incident raises major issues in the area of chemical safety and response to chemical disasters. Given the large size of the chemical industry and the production, transport and distribution of large number and quantity of chemicals, this concern should be addressed. Local authorities which are the first responders in a chemical accident need specialized technical and institutional support to deal with such situations. Most disaster management agencies in countries have built capacity and to operational guidelines deal with chemical accidents. The requirements also include post incident decontamination.

The International Programme on Chemical Safety (IPCS) led by the WHO was formed in 1980, jointly with the International Labour Organization and the United Nations Environment

Programme, to establish a scientific basis for safe use of chemicals and to strengthen national capabilities and capacities for chemical safety. The IPCS also covers dangerous chemicals in the environment and measures to deal with them.

## Environmental Health

Air and water pollution especially in urban areas has become a growing health problem. Air quality in major cities in India, China and other developing countries has deteriorated, exposing large populations to the effects. The growth of cities and economic activities such as transportation and industry add to the problem. The issue is how to improve air quality while maintaining economic development. Similarly water in rivers and underground is becoming highly polluted by discharges of industrial and municipal effluents, compromising access to safe drinking water for large populations. These problems are likely to severely impact human health during the current century and lead to sharp rise in diseases such as cancer and Alzheimer's, adding to the already heavy health care burden of developing countries.

Tackling this problem requires concerted national action among various agencies, as well as international cooperation. Air and water borne pollution can go across national boundaries. For example PM1 sized particles can travel several thousand kilometers in the air. These problems are now attracting increasing public concern and official agencies are taking some measures to deal with them. But far more needs to be done. Technology for cleaner transport, production, and municipal waste management plays a key role in this field and should be the focus of internationally supported R & D efforts.

## Food Safety

In recent years, food products have been found to contain harmful chemicals. Safety standards have been introduced in the EU and other countries covering the presence of such chemicals. Agrochemicals are used in agricultural practices and animal husbandry with the intent to increase crops and reduce costs. Such agents include pesticides (e.g., insecticides, herbicides, rodenticides), plant growth regulators, veterinary drugs (e.g., nitrofuran, fluoroquinolones, malachite green, chloramphenicol), and bovine somatotropin (rBST). Environmental contaminants are chemicals that are present in the environment in which the food is grown, harvested, transported, stored, packaged, processed, and consumed. The physical contact of the food with its environment results in its contamination. The limits on concentrations of contaminants allowed by legislation are often well below toxicological tolerance levels, because such levels can often be reasonably achieved by using good agricultural and manufacturing practices. The impact of chemical contaminants on consumer health and well-being is often apparent only after many years of processing. Prolonged exposure at low levels may cause cancer or have mutagenic or teratogenic effects. Chemical contaminants present in foods are often unaffected by thermal processing. Workers in the agriculture and food processing sectors may get exposed to much higher levels of chemicals due to their work environment.

There are several cases of banned pesticides or carcinogens found in foods. Greenpeace exposed in 2006 in China that 25% of surveyed supermarkets agricultural products contained banned pesticides. In India, soft drinks were found contaminated with high levels of pesticides and insecticides, including lindane, DDT, malathion and chlorpyrifos[8]. Formaldehyde, a carcinogen was found in Vietnamese national dish, Pho, in 2007. In Indonesia (2005), carcinogenic formaldehyde was found added as a preservative to noodles, tofu, salted fish, and meatballs.

Food safety requires an effective testing and monitoring system to detect harmful chemicals and take appropriate action. Cleaner technology in agriculture, animal husbandry, food processing etc. is also important to deal with this issue. Further, the international trade in agricultural and food products could be affected by disagreement over standards for food safety. There is need to harmonize the standards set for presence of harmful chemicals in food and agricultural products[9], through a fair and transparent process, using mechanisms such as the Codex Alimentarius Commission (CAC). These matters need international cooperation.

## Human Reproduction, Genomics, and related issues

Technological advances have now made it possible to fertilize human eggs outside the body and implant them successfully in a recipient female uterus, and thereby produce offspring who owe their existence to two biological parents and one surrogate mother. This has enabled many couples to achieve their dreams of having children. However a number of ethical and regulatory issues have arisen over the roles of the biological parents and the surrogate mother, in order to safeguard the rights and prevent abuse of any of the parties. In this field technology has moved beyond the social framework to handle such questions. The use of embryonic stem cells for research has become a subject of controversy and countries differ over regulation of such research even though there are many potential benefits.

Similarly, the availability of rapid genome sequencing could make it possible for misuse of such data through discrimination against those whose genomes may give them a higher risk of disease or other condition. Moving beyond, in the event that genetic modification of an embryo becomes technically feasible, there would be many ethical questions over regulating such research and development. Advances in biotechnology may soon make it possible to generate replacement human tissues and even whole organs to replace diseased ones and thereby save lives and extend life spans. Should such technology be made widely available at affordable prices or exploited for commercial benefits of a few? These challenges are likely to arise in the 21 st century.

The human health sector involves a variety of issues where international cooperation in research and technology plays an important role. Indeed without strong cooperation on this front, problems will be difficult to solve. Achieving the UN's Suasainable Development Goal 3 which reads "Ensure healthy lives and promote well-being for all at all ages" will require tremendous efforts. In almost all areas of human health, much more will need to be done to attain the 10 targets set for 2030.

## Endnotes

1  Origin and development of health cooperation, WHO, http://www. who.int/global_health_histories/background/en , accessed 3-1-2017

2  In China after the two Opium Wars of the 19 th century, in which Britain defeated China, severe limitations were placed upon Chinese sovereignty through the treaties that ended the wars. These restrictions prevented the Chinese government from stopping the continual flow of opium imports from India – an important source of revenue for the British colonial government in India. Britain justified the trade with arguments that if it were to stop, the British India market share would immediately be absorbed by other producers such as Persia and Turkey, and China would simply increase its domestic opium production.

3  The History and Development of the Leading International Drug Control Conventions, Jay Sinha, 2001, Parliament of Canada, http:// www.parl.gc.ca/content/sen/committee/371/ille/library/historye.htm ,accessed 6-1-2017

4  N-bomb was discovered in 2003 by chemist Ralf Heim at the Free University of Berlin, Germany. It was derived from a group of drugs called the 2C family of phenethylamines (PEA). See The truth about synthetic drugs , http://www.drugfreeworld.org/drugfacts/synthetic. html , accessed 9-1-2017

5  N-Bomb: The new drug on Narcotics Control Bureau's radar, Rohit Alok, Indian Express, 8 Aug 2016, http://indianexpress.com/article/ india/india-news-india/n-bomb-drug-mumbai-goa-narcotics-ncb-lsd-new-illegal-2960458 , accessed 9-1-2017

6   TRIPS and pharmaceutical patents, WTO Fact Sheet, September 2006, https://www.wto.org/ENGLISH/tratop_e/trips_e/tripsfactsheet_ pharma_2006_e.pdf , accessed 9-1-2017

7   Global AIDS update, UNAIDS, 2016, http://www.unaids.org/sites/ default/files/media_asset/global-AIDS-update-2016_en.pdf , accessed 9-1-2017

8   Supreme Court reminder to FSSAI on monitoring pesticides in food commodities, Down to Earth, 24 October 2013, http://www. downtoearth.org.in/news/supreme-court-reminder-to-fssai-on-monitoring-pesticides-in-food-commodities-42537 ,accessed 10-1-2017

9   Countries with a high percentage of total imports in agriculture tend to prefer stricter maximum residue limits (MRL). Countries with greater employment in agriculture and countries that produce the regulated commodities tend to have looser MRLs. Higher incomes and public healthexpenditures are also associated with stricter MRLs.

## Chapter 13

# Intellectual Property Rights – Converting Knowledge into Economic Power

*"The patent system .... secured to the inventor, for a limited time, the exclusive use of his invention; and thereby added the fuel of interest to the fire of genius, in the discovery and production of new and useful things."*

*– Abraham Lincoln, 1859*

### Introduction

The importance of intellectual property arises from its stimulation of robust research, innovation and development and the consequent development of a knowledge economy. This is the foundation for competitiveness of an economy in the modern post industrial age. Therefore the IPR ecosystem in a country assumes prime importance.

The concept of intellectual property (IP) evolved since the 18th century as a means of giving a monopoly for the use of inventions, and other creative works. In modern usage, intellectual property covers two broad categories – copyrights and related rights, and industrial property. The latter includes patents, industrial designs, trade secrets, trade dress and plant breeders rights, which are the results of innovation, and trademarks and geographical indications, which distinctly identify a product or service[1].

The main purpose of giving intellectual property rights to the creators is to encourage and reward innovation and creativity and enable them to benefit from the exploitation of the IP through the market, and thereby promote human progress and knowledge development. According to one definition, the purpose of IPRs is (1) to give statutory expression to the moral and economic rights of creators in their creations and the rights of the public in access to those creations; and (2) to promote, as a deliberate act of Government policy, creativity and the dissemination and application of its results and to encourage fair trading which would contribute to economic and social development. With the advancement of science and technology and the high benefits arising from new discoveries, IPRs have assumed very great importance in the world. Essentially, possession of IPRs is a critical means of gaining dominance in the market through knowledge.

## The IPR Framework

Over the years, several international treaties have been formulated and adopted giving recognition to IPRs, and are administered by the World Intellectual Property Organization (WIPO). Copyrights and related rights are protected under the Berne Convention of 1886 (the latest amendment in 1979), which recognizes that a copyright exists the moment a work is published rather than requiring registration. It also imposes a requirement that all signatory countries recognize copyrights held by the citizens of any of its parties. It requires member states to provide strong minimum standards for copyright protection. Copyrights are for 50 years, except for photographic or cinematographic works. The WIPO Copyright Treaty (WCT) of 1996 deals with the protection of works and the rights of their authors in the digital environment. In addition to the rights recognized by the Berne Convention, they are granted certain economic rights. This Treaty also covers (i) computer programs, whatever the mode or form of their expression; and (ii) compilations of data or other material ("databases").

The Paris Convention, adopted in 1883 (latest amendment in 1979) covers industrial property in the widest sense, including patents, trademarks, industrial designs, utility models, service marks, trade names, geographical indications and the repression of unfair competition. It was one of the first intellectual property treaties. It established a Union for the protection of industrial property. Applicants from a country are entitled to national treatment in any other member state. The date of first filing of an application is the effective date of filing in another member state.

IPRs are enforced through national legislation which is a requirement under the international treaties. Violation of intellectual property rights, called "infringement" with respect to patents, copyright, and trademarks, and "misappropriation" with respect to trade secrets, may be a breach of civil law or criminal law, depending on the type of intellectual property involved, jurisdiction, and the nature of the action. In 2011 trade in counterfeit copyrighted and trademarked works was estimated to be worth $600 billion worldwide and accounted for 5–7% of global trade. China in particular has attracted concern over copyright and intellectual property violations especially in the automotive and electronics industries. However in recent years infringements have reduced as implementation of laws has become more effective.

## Balancing Incentives and Public Interests

The system of IPRs has been criticized on several grounds. Critics of intellectual property, point at IPRs as harming health in the case of pharmaceutical products, preventing progress, and benefiting monopoly interests to the detriment of the public. The public interest may be harmed by ever-expansive monopolies in the form of copyright extensions, software patents, and business method patents. More recently scientists and engineers are expressing concern that patent thickets are undermining technological development even in high-tech fields like nanotechnology. Intellectual property rights to early generations of inventors may discourage subsequent innovation.

Policies that encourage the diffusion of ideas and modify patent laws to facilitate entry and encourage competition may be more effective in encouraging innovation. The granting of product patents may in fact discourage research into alternative cheaper pathways to manufacture the product and limit the growth of human knowledge, and discourage competition.

The pharma industry has been the centre of controversy related to abuse of market power provided through strong IPRs. Instances of refusal to manufacture products locally and force imports may lead to unfairly high drug prices. The granting of stronger IPR to pharma companies has in general led to higher prices of drugs even those needed in the public interest. Therefore the IPR system does provide some avenues for balancing public interest through compulsory licensing. An IPR driven regime may also not be conductive to the investment of R&D in products that are socially valuable to predominately poor populations, such as drugs and vaccines for treating diseases prevalent in the poorer countries.

## TRIPs and IPRs

In the 1980s leading multinational pharma companies, concerned at growing competition from lower cost manufacturing in developing countries such as India, Brazil, etc. mounted a strong lobbying effort in the US to make IPRs a major issue in trade policy. The US a world leader in terms of IPRs, together with the European Union, Japan and other IPR rich developed nations mounted a strong campaign for inclusion of IPR related agreement under the Uruguay Round negotiations of GATT ( later to become the WTO) which had begun in 1986. Pressures including unilateral economic encouragement under the Generalized System of Preferences and coercion under Section 301 of the US Trade Act played an important role in defeating opposition from developing countries such as India, Brazil, Thailand, and Caribbean Basin states.

The final Agreement on Trade-Related Aspects of Intellectual Property Rights (TRIPS), 1994, sets down minimum standards for many forms of intellectual property (IP) regulation as applied to nationals of other WTO Members. Because ratification of TRIPS is a compulsory requirement of World Trade Organization membership, any country seeking to obtain benefits of WTO membership must enact the strict intellectual property laws mandated by TRIPS which became the most important multilateral instrument for the globalization of intellectual property laws. Unlike other agreements on intellectual property, TRIPS has a powerful enforcement mechanism. States can be disciplined through the WTO's dispute settlement mechanism.

The TRIPS agreement introduced intellectual property law into the international trading system for the first time and remains the most comprehensive international agreement on intellectual property to date. In 2001, developing countries, supported by civil society groups, concerned over the application of TRIPS and its impact on availability of life saving drugs initiated a round of talks that resulted in the Doha Declaration. The Doha declaration is a WTO statement that clarifies the scope of TRIPS, stating for example that TRIPS can and should be interpreted in light of the goal "to promote access to medicines for all."

TRIPS requires WTO members to provide copyright rights, covering content producers including performers, producers of sound recordings and broadcasting organizations; geographical indications, including appellations of origin; industrial designs; integrated circuit layout-designs; patents; new plant varieties; trademarks; trade dress; and undisclosed or confidential information. TRIPS also specifies enforcement procedures, remedies, and dispute resolution procedures. Protection and enforcement of all intellectual property rights shall meet the objectives to contribute to the promotion of technological innovation and to the transfer and dissemination of technology, to the mutual advantage of producers and users of technological knowledge and in a manner conducive to social and economic welfare, and to a balance of rights and obligations. Developing countries were allowed extra time till 2005 to become TRIPs compliant while for least developed countries the deadline was extended up to 2013, and until 1 January 2016 for pharmaceutical patents, with the possibility of further extension.

TRIPS has led to some controversies. The most visible conflict has been over availability of lower priced AIDS drugs in Africa in the interest of public health, which led to the Doha Declaration. Another issue concerns the granting of software and business method patents. Many developing countries have not made full use of TRIPS flexibilities (compulsory licensing, parallel importation, limits on data protection, use of broad research and other exceptions to patentability, etc.) to the extent authorized under Doha.

Criticism of TRIPs includes the wealth transfer from people in developing countries to copyright and patent owners in developed countries, and failure to accelerate investment and technology flows to low-income countries. Lengthy patent periods under TRIPs have unduly slowed the entry of generic substitutes and competition to the market. The illegality of pre-clinical trials or submission of samples for approval until a patent expires have been blamed for driving the growth of a few multinationals, rather than developing country producers. The importance of TRIPS in the process of generation and diffusion of knowledge and innovation has been overestimated.

## India and TRIPs

India had a key role in the TRIPs negotiations. The original Indian Patent Act had not recognized product patents, but only process patents. Thus Indian companies could take advantage of this and carry our research to develop viable alternative processes to manufacture useful molecules for medical use. This "reverse engineering" led to Indian companies being able to manufacture drugs at far lower cost that the original patent holders. The heavy expenditure of clinical trials and approvals for the drugs was also not a burden. Indian companies were seen as a serious threat by the world's leading pharma producers, especially in the US. This led to a strong reaction and lobbying efforts by the latter to include strong protection of IPRs in the Uruguay Round trade negotiations.

India and several other developing countries opposed the position taken by the US and some other developed countries. However the skilful application of threats and incentives caused divisions in and defections from the developing countries group, with India getting more and more isolated. The US then threatened to use trade sanctions against India if it did not give ground. Finally, India had to yield on the question of product patents, despite the support of some civil society groups which saw India as the pharmacy of the developing world. India and Brazil were able to supply anti AIDs drugs at a fraction of the cost of supplies from the established pharma companies. In the final TRIPs agreement product patents were recognized and India had to comply with the new agreement by 2005. The changes in the Indian Patents Act were finally made in the face considerable domestic opposition. Over the period from 1995 onwards, the prices of pharma products in India began to rise steadily as a consequence of less competition. Indian pharma companies have adjusted to the situation by taking on more research to develop new molecules, and by taking up manufacture of generics and drugs whose original patents have expired.

Some breathing space was secured in the Doha Declaration of 2001. This was mainly due to NGO action in the US and Europe and from African countries affected by AIDs which needed affordable anti-AIDs drugs. Flexibility was allowed in TRIPs under this Declaration in the case of public health emergencies. But this flexibility was not included as part of the TRIPs agreement itself.

## Geographical Indications

Geographical indications (GIs) may have economic value because it may act as a certification that the product possesses certain qualities, is made according to traditional methods, or enjoys a certain reputation, due to its geographical origin. The TRIPS agreement strengthens protection of GIs. Governments must provide legal opportunities in their own laws for the owner of a GI registered in that country to prevent the use of marks that mislead the public as to the geographical origin of the good. India has also had to deal with disputes over the GI "Basmati" for long grain rice which was settled in 2016 in favour of rice grown in seven Indian states. The claim had been challenged by rice growers in Lahore, Pakistan, as well as internally by other Indian states. Some 26 other Indian well known products have also had their GIs registered.

## Copyrights Protection and India

Creative works by authors such as books, writings, movies, music, etc. are protected under national laws and an international treaty, the Berne Convention 1886 and its subsequent amendments (latest being in 1979). Copyright is a legal right granted to the creator of an original work covering exclusive rights for its use and distribution. These rights frequently include reproduction, control over derivative works, distribution, public performance, and "moral rights" such as attribution. The duration of a copyright spans the author's life plus 50 to 100 years . Most countries recognize copyright in any completed work, without formal registration.

The development of digital media and computer network technologies have introduced new difficulties in enforcing copyright, and given rise to additional challenges to the basis of copyright law. Piracy of music and movies has become a challenge, as more and more use is made of electronic formats rather than physical books, recordings and films. Businesses with great economic dependence upon copyright, such as those in the music business, have advocated the extension and expansion of copyright and sought additional legal and technological enforcement.

The World Intellectual Property Organization Copyright Treaty (WIPO Copyright Treaty or WCT), provides additional protections due to advances in information technology. Computer programs are protected as literary works, and that the arrangement and selection of material in databases is protected. It provides authors of works with control over their rental and distribution beyond the Berne Convention. It also prohibits circumvention of technological measures for the protection of works (anti copying security measures) and unauthorized modification of rights management information (such as region specific video discs) contained in works. As of 2016, the treaty has been ratified by 94 states.

India is a major producer of music, movies and has a large software and related services industry. It therefore has a strong interest in copyright protection. India enacted the Copyright Act 1957 and the law has been amended six times since 1957, the last being in 2012. However India has not joined the WCT, but its law is in conformity with the WCT. India has a big stake in enforcing copyright protection and fighting piracy and online copyright violation which has become a serious issue for Indian entertainment industry. The owners of Indian films and music records have adopted many measures to prevent online theft of their movies and songs but with little success so far. Indian law enforcement agencies have expressed their inability to locate the person uploading the copyright violating content due to use of anonymous services. Therefore securing greater international cooperation to curb copyright violations is a challenge for India.

## Abuses of Patent Protection

In recent years, there have been abuses of the system of granting patents. A U.S. patent on turmeric was granted to the University of Mississippi Medical Center in 1995, for the "use of turmeric in wound healing." This patent also granted them the exclusive right to sell and distribute turmeric. Two years later, a complaint was filed by India's Council of Scientific and Industrial Research, which challenged the novelty of the University's "discovery," and the U.S. patent office investigated the validity of this patent. In India, where turmeric has been used medicinally for thousands of years, concerns grew about the economically and socially damaging impact of this legal "biopiracy." In 1997, the patent was revoked. Turmeric for treating wounds had in fact been traditionally practiced in India for thousands of years, as was eventually proven by ancient Sanskrit writings that documented turmeric's extensive and varied use throughout India's history. The US Department of Agriculture (USDA) and the US chemical major W.R. Grace were granted a patent in 1995 by the European Patent Office on a fungicide formation from the seeds of Neem. This was challenged and after a ten year long battle the patent was refused in 2005. This problem of biopiracy needs to be combated by developing countries.

Another controversial issue is the grant of biological patents which may include biological technology and products, genetically modified organisms and genetic material. The patenting of genes is a controversial issue, because it treats life as a commodity, or undermines the dignity of people and animals by allowing ownership of genes. Some say that living materials occur naturally, and therefore cannot be patented. Gene patents may inhibit access to genetic testing for patients and hinder research on genetic disease. About 3,000 to 5,000 patents on human genes have been granted in the United States. Although gene patents often base their claims at least partly on whole genes, they also cover many kinds of inventions involving the components of genes and genetic technologies. In 2013, the US Supreme Court issued a ruling that bans the patenting of naturally occurring genes but allows edited or artificially created DNA to be patented. The ruling partially clarified the situation, but left many questions unanswered.

### Innovation and Research

IPRs play a key role in driving research, innovation and development of technology. The ecosystem in which IPRs operate contains many types of players. The research workers in centres in private and public institutions are the originators of innovation and knowledge. This feeds into another group of institutions which support the commercialization of new knowledge through finance, market research and feedback, business support structures, and startup enterprises using new knowledge. Governments can and should play a key role in providing various incentives for all the types of activities such as research, business development, and marketing.

The advantage of scientific talent in many countries can be exploited by foreign institutions through attracting them to the advanced countries where the ecosystems are fully mature and offer the attraction of high salaries, However, in the US, due to various factors, there are pressures to create barriers to the hiring of so called tech workers which have gained political traction. Also the bureaucratic approach to security clearances for technology workers from foreign countries has created barriers. These counterproductive measures create incentives for multinational companies to locate their research centres abroad where conditions for mobility of researchers are easier. It may also result in large scale migration of science and technology talent to countries such as China[2] which are rapidly increasing R & D spending, provided a conducive environment[3] is available for such migration.

An alternative approach would be to simply exploit the talent through their own institutions in the country where the researchers are located, with relatively lower salaries, and reaping the global benefits of new IPRs and innovations resulting from the research. Another approach would be to acquire promising start ups at lower cost before they become fully operational. Foreign companies such as GE have followed such an approach by setting up large research centres in talent rich countries such as Israel, Russia, and India. In such a model, the largest portion of the value addition arising from commercial development of an innovation accrues to the foreign entity, while the research worker may get a relatively smaller part of the benefit, depending on the patenting conditions and the business agreements. To get the best benefits out of talented researchers is therefore a challenge, requiring a strong and mature research and IPR ecosystem so that the highest benefits accrue to the country and the research worker.

All countries are competing in the race to benefit from research and innovation. Government actions may play an important part in determining success. India for example has embarked on measures to improve ease of setting up and doing business, encouraging start ups, etc.

## Going beyond TRIPs – the TPP

After the WTO TRIPs agreement, the US in particular, being a leader in terms of IPRs has sought to leverage its strength by securing increasingly stronger and wider IPR protection. US business and government have viewed this strategy as key to economic dominance and prosperity in a globalizing world. More aggressive IPR provisions have therefore been negotiated in US bilateral trade agreements with several countries such as Australia, Chile, Peru, Jordan, Singapore, Vietnam, Morocco and Bahrain. The draft Trans Pacific Partnership (TPP) agreement chapter on IPRs contains provisions[4] that go well beyond TRIPs in several key aspects. Some of these provisions are – (1) Trademark terms of protection increased from 7 to 10 years and the removal of barriers for the protection of sound marks; (2) Copyright duration of protection increased from 50 to 70 years and stronger copyright enforcement, including the possibility of criminal prosecution against acts of removal of rights management information and the requirement that TPP countries be signatories of WIPO "Internet treaties"; (3) Requirement of enforceable legal means for the protection of trade secrets (TRIPS does not specify these means); (4) Protection of undisclosed test data submitted for marketing approvals (at least 10 years in the case of agricultural chemicals and 5 to 8 years in the case of pharmaceuticals; TRIPS does not have such a requirement); and (5) Explicit protections for new pharmaceutical products that are or contain a biologic (the TPP is the first trade agreement to do this).

The TPP has been opposed during the 2016 US elections by both sides, and it is unlikely to be finalized quickly. However, the opponents of the TPP have argued that the US gave away too many concessions in the TPP. Therefore pressures to introduce the TPP and TPP plus provisions in future bilateral trade agreements are likely to increase, given the perception that the US can leverage its influence better in bilateral negotiations and extract more concessions from its partners, as well as the strong support from the US business sector.

The TPP may be dormant, but its spirit may live on through such negotiations in future.

## Future Challenges in IPR

As mentioned, IPR challenges arise on several fronts. Nationally the importance of a strong ecosystem to include research, innovation and IPRs cannot be overemphasized. Failure to do this will result in the advantages of scientific talent and innovation being drained away by foreign companies. As knowledge advances, efforts by IPR leading countries to strengthen the IPR regime to extract maximum advantage will continue. This may take the form of bilateral trade and economic partnership agreements. National capacity to deal with such pressures will be critical. Advances in technology will bring up new aspects of IPR, for example the move to patent genes occurring naturally vs patenting of synthetic genes, patenting of surgical and medical procedures, etc. Reviving the WTO as the most important global forum for all trade related issues will be an important future challenge.

The importance of research and innovation cannot be overemphasized. Under the UN's Sustainable Development Goal No 9 for 2030, one of the targets reads "Enhance scientific research, upgrade the technological capabilities of industrial sectors in all countries, in particular developing countries, including, by 2030, encouraging innovation and substantially increasing the number of research and development workers per 1 million people and public and private research and development spending."

## Endnotes

1 What is Intellectual Property? WIPO, www.wipo.int/edocs/pubdocs/ en/intproperty/450/wipo_pub_450.pdf , accessed 10-1-2017

2 China Plans First Immigration Agency to Lure Overseas Talent, Bloomberg news, 19 July 2016, https://www.bloomberg.com/news/

articles/2016-07-18/china-said-to-create-new-office-to-lure-overseas-work-talent, accessed 17-4-2017

3   China dangles green cards to entice foreign science talent, M. Hvistendahl, Science, 27 Jan 2015, http://www.sciencemag.org/news/2015/01/china-dangles-green-cards-entice-foreign-science-talent, accessed 17-4-2017

4   TPP: The New Gold Standard for Intellectual Property Protection in Trade Agreements?, Prima Braga, C.A., Huffington Post, 24 March 2016, http://www.huffingtonpost.com/eastwest-center/tpp-the-new-gold-standard_b_9544428.html, accessed 10-1-2017

# Chapter 14

# Technology Control – Preventing Spread of Weapons

*"The United States has a profound interest in keeping weapons of mass destruction and other sophisticated weaponry out of unstable regions and away from rogue states and terrorists. In the twenty first century, many of the threats to our security will come not from great power conflict but from states that defy the international community and violent groups seeking to undermine peace, stability and democracy."*

*– President Clinton, speech on 24 October 1997.*

## Introduction

The main motive for technology control has historically been to deny opponents the benefits of new technology. The perceived benefits include military capability as well as economic rise to challenge the prevailing order. Even in peace time, or in non-conflict situations, the denial or restriction regimes may be used to protect against competition from rivals and to maintain market dominance. Such dominance can often be leveraged into higher profits. However, the denial of strategic technology may result in the rival embarking on its own research effort which may eventually generate alternative products and systems of lower cost and greater effectiveness. Controls therefore are double edged weapon, beyond a point it can stimulate the enemy to greater efforts to equalize the advantage or surpass it. Such was the case with the Soviet Union in the case of the thermonuclear bomb and space flight.

The opposite side of control is technology transfer which has been sought be developing countries, especially in critical areas such as agriculture, food, and health care. Technology is usually developed by private companies and transfer of such technology requires agreed technology transfer arrangements or licensing, which may be used to extract profits from transfer of technology. The patent system which gives a monopoly over new technology also helps by reducing competition. Developing countries' efforts to negotiate a binding international code of conduct for technology transfer were effectively blocked by the US and OECD countries which did not want any international regulation of private entities and governments to negotiate technology transfer arrangements. This kept the door open to restrictive and monopolistic business practices, such as imposing restrictions on production, marketing, and product development. In strategic as well as economic areas, governments especially of the advanced countries have been involved in control of technology through various national and international arrangements which this chapter reviews.

## Controls over Technology

The fact that technology development results in vastly more effective and destructive weapons has been well appreciated. Countries which have such technology and weapons have a military advantage, which they may seek to perpetuate through controls and restrictions over technology through national means or in cooperation with allies. The rivals sought to gain access to military technology and information through espionage, smuggling etc. This resulted in national mechanisms to ensure that sensitive technology would not be made accessible directly or indirectly to rivals. Of course, rivals would also embark on their own efforts at research and development of technology which had strategic importance, such as nuclear or missile technology.

In the US, a leading nation in technology development, an elaborate system of export controls had been built up to prevent exports of sensitive products and technology to rivals. During the Cold War period this gained considerable importance, as the US sought to preserve its technology advance over the Soviet bloc. Starting with the Trading with Enemy Act of 1917, the US introduced the Export Control Act of 1949 (replaced by the Export Administration Act in 1979); to restrict sensitive exports to the Soviet Bloc and generally to support US security and foreign policy objectives. A large number of US agencies especially the Commerce, State and Defence departments, became involved in implementing the controls over exports, coordinated by the Bureau of Industry and Security of the Department of Commerce. Lists of entities deemed to be working against US interests were maintained which would be subject to export restrictions.

In order to close off leakage of sensitive technology from other advanced countries allied to the US, international mechanisms were negotiated to seal this gap. An informal body, the Coordinating Committee for Multilateral Export Controls (CoCom) was established by Western bloc powers led by the US, during 1945-49, to put an arms and related technology embargo on COMECON countries. It established lists of items to be restricted which were continually revised, as East West relations waxed and waned. An important source of tension was the conflict between CoCom restrictions and the interests of some member states to increase their trade with Comecon countries. CoCom ceased to function in 1994, at the end of the Cold War, and the Wassenaar Arrangement, was established as its successor. Despite its informal status and the increasingly divergent interests of its members, it had functioned for nearly 44 years, largely due to the great influence of the US. The CoCom arrangements were basically arrangements negotiated largely by the US under the fig leaf of a multilateral entity. The US government made it clear that it would exercise control over exports by any foreign company using sensitive technology from the US. Punitive measures could include refusal of further US supplies or fines and penalties against the foreign company's presence in the US.

In the case of India, supplies of high technology items from US companies were restricted until India, seen as an ally of the Soviet Union negotiated and signed a MoU and an implementation agreement with the US in 1984 clearing the way for Indian access to civilian and dual-use technologies and some military assistance. After this a large number of items of advanced technology items were released. But even then, the US refused to supply the advanced Cray supercomputer in 1987 and offered a less powerful substitute instead[1]. The use of US computers for a project in Syria implemented by the Indian public enterprise CMC was also subject to US licensing approval. Pressure could be exerted by delays in granting approvals or by advising the US supplier to withdraw offers.

In recent years, India-US relations have improved considerably. The US strongly supports India's inclusion in all the four multilateral technology control regimes – the Nuclear Suppliers Group (NSG), the Missile Technology Control Regime (MTCR), the Australia Group, and the Wassenaar Arrangement (these control regimes are described in detail below). The growing relationship has also eased some of the export control issues between the two sides, especially in defence related technology. However, India being a non-NATO country there are restriction on visas for Indian scientists to participate in research work in the US especially in sensitive areas.

## Nuclear Technology – the NSG

The most important of the technology control regimes is the Nuclear Suppliers Group (NSG) set up in order to reinforce the Nuclear Nonproliferation Treaty. After India, a non-signatory to the NPT conducted an underground nuclear explosion in 1974. This demonstrated that certain non-weapons specific nuclear technology could be readily turned to weapons development. The Nuclear Non-Proliferation Treaty (NPT) members saw the need to further limit the export of nuclear equipment, materials or technology. Another purpose was to bring in non-NPT and non-Zangger Committee[2] nations, then specifically France[3], with significant nuclear capability into the regime. A series of meetings in London from 1975 to 1978 resulted in agreements on the guidelines for export; these were published by the International Atomic Energy Agency (IAEA) as INFCIRC/254 (essentially the Zangger "Trigger List"). Listed items could only be exported to non-nuclear states if certain IAEA safeguards were agreed to or if exceptional circumstances relating to safety existed. This subject was largely handled under the IAEA framework until the revelations about the Iraqi weapons program following the first Gulf War led to a renewed activity aimed at tightening[4] of the export of so-called dual-use equipment. At the first meeting since 1978, held at The Hague in March 1991, the twenty-six participating governments agreed to the additions, which were published as the "Dual-use List" in 1992, and also to the updating of the original list to more closely match the most recent Zangger list. The NSG reaches decisions by consensus. Membership of the NSG has increased from the initial 7 to 48 at present. Several former Soviet republics also became members. China became a member in 2004. The European Commission and the Zangger Committee Chair participate as observers. India and Pakistan, non-NPT signatories and Namibia, are candidates for membership.

India's candidacy for membership of the NSG has been supported strongly by the US, which since 2005 has led the move to remove India's isolation in nuclear commerce. The waiver from the NSG secured under much pressure from the US was a key step in enabling India to have normal nuclear commerce, including fuel supplies and reactors for its ambitious civil nuclear programme. India has indicated that it would abide by the same conditions as and support the NPT as a nuclear weapons state party, and has put in place NSG compliant export control regimes. However several countries including China have opposed India's membership on the grounds that it was not an NPT signatory, although in China's case strategic considerations also have played a part. As a result of the obstructionist position taken by some NSG members, support in India for joining the NSG has become weaker and there is questioning about whether NSG membership is worth the political capital expended. The NSG Chair, Argentina has been tasked to find a constructive solution.

Has the NSG been successful in its main objective of curbing nuclear weapons proliferation? Certainly the export control measures have made is much more difficult for non-NPT states to develop nuclear capability. However, India developed its civil and strategic nuclear capacity by its own indigenous efforts, while Pakistan did so using technology and support from China. Israel, a non NPT state, has developed a substantial arsenal of nuclear weapons as a deterrent against hostile neighbours. Iraq, Libya, Iran and North Korea despite the NPT have developed nuclear capability, the first three using material and support from other countries or smuggling operations such as the A Q Khan network. North Korea withdrew from the NPT and carried out nuclear tests in a drive for security in the face of US hostility and determination to change the regime. The nuclear deal framework and the Joint Comprehensive Plan of Action of 2015 between Iran and the P5+1[5]countries delays but does not eliminate Iran's capability to build a nuclear weapon in future should it decide to do so. The fragility of this deal is underlined by the past history of US-North Korea nuclear agreement of 1994 which fell through. Thus it could be argued that the NSG is largely a technical effort, and does not address the root cause of nuclear proliferation, i.e. the security and existential threats perceived by certain countries that compels them to pursue strategic nuclear capability.

## Missile Technology – The MTCR

The Missile Technology Control Regime (MTCR) established in 1987 by the G7 countries in order to curb the spread of unmanned delivery systems for nuclear weapons, specifically delivery systems that could carry a payload of 500 kg for a distance of 300 km. In 1992, it was agreed to expand the scope of the MTCR to include nonproliferation of unmanned aerial vehicles (UAVs) for all weapons of mass destruction. Prohibited materials are divided into two Categories, which are outlined in the MTCR Equipment, Software, and Technology Annex. Membership has grown to 35 nations. India also joined in June 2016 adhering to the MTCR Guidelines unilaterally. In 2004 China applied to join the MTCR, but members did not offer China membership because of concerns about China's export control standards. Israel, Romania and Slovakia have also agreed to voluntarily follow MTCR export rules even though not yet members. Pakistan has an ongoing dialogue with the MTCR Chair about membership.

The MTCR has not been able to limit the spread of missile capability. China, Iran, Israel, North Korea, and Pakistan continue to advance their missile programs, including medium-range ballistic missiles that can travel more than 1,000 kilometers and are developing missiles with much greater ranges. Israel and China in particular have already deployed strategic nuclear SLCMs and ICBMs and satellite launch systems. Non MTCR members such as North Korea, have become a source of ballistic missile proliferation. China has supplied ballistic missiles and technology to Pakistan and Saudi Arabia, Iran has supplied missile technology to Syria. In 2002, the MTCR was supplemented by the International Code of Conduct against Ballistic Missile Proliferation (ICOC), also known as the Hague Code of Conduct, which calls for restraint and care in the proliferation of ballistic missile systems capable of delivering weapons of mass destruction, and has 119 members, thus working parallel to the MTCR with less specific restrictions but with a wider membership.

## Chemical Technology – the Australia Group

The Australia Group is an informal group of countries and the European Commission, established in 1985, following the use of chemical weapons by Iraq in 1984, to control exports which might contribute to the spread of chemical and biological weapons. The group membership has increased from the initial 15 members in 1989 to 42 members at present. Australia manages the secretariat.

Myanmar, China, India, Vietnam, the Philippines and Singapore are non-member dialogue partners.

Members of the group maintain export controls on a uniform list of 54 compounds, including several that are not prohibited for export under the Chemical Weapons Convention, but can be used in the manufacture of chemical weapons. In 2002, the group took further steps to strengthen export control. Export to another state that had already been denied an export by any other member of the group requires prior consultations with that member state. Member states are required to halt all exports that could be used by importers in chemical or biological weapons programs.

The Group is an informal association that works on the basis of consensus. It aims to allow exporters or transshipment countries to minimize the risk of further proliferation of chemical and biological weapons (CBW). The Group meets annually to assess ways in which the national level export licensing measures of its participants can collectively be rendered more effective to ensure that would-be proliferators are unable to obtain necessary inputs for CBW programs, which are banned under the provisions of the Chemical Weapons Convention (CWC). All States participating in the AG are Parties to the CWC and the Biological and Toxin Weapons Convention (BTWC). The AG aims to limit the spread of CBW through the control of chemical precursors, CBW equipment, and BW agents and organisms. All participating countries have licensing measures covering over 60 CW precursors. Participating countries also require licenses for the export of 1) dual-use chemical manufacturing facilities and equipment and related technology; 2) plant pathogens; 3) animal pathogens; 4) biological agents; and 5) dual-use biological equipment.

India has a well developed chemical industry and a comprehensive export controls system for chemical agents that can be used for chemical warfare. India is also a member of the CWC and the BWC. The US has supported India's candidature for the Australia Group. India has set up a system of licensing for exports on the Special Chemicals, Organisms, Materials, Equipment, and Technologies (SCOMET) list. This list includes material related to Weapons of Mass Destruction Technology. However, the SCOMET list is not fully consistent with the Australia Group's Chemical Dual Use List. There is an ongoing dialogue between India and the Australia Group over these issues. Some also suggest that India would like to join the Australia group once the membership in the NSG and the Wassenaar Arrangement is resolved satisfactorily.

## Dual use Technology – The Wassenaar Arrangement

The Wassenaar Arrangement (WA) on Export Controls for Conventional Arms and Dual-Use Goods and Technologies, known as the Wassenaar Arrangement, is the successor to the Cold War-era Coordinating Committee for Multilateral Export Controls (COCOM), and was established in 1996, in Wassenaar, the Netherlands, near The Hague. The Wassenaar Arrangement is more liberal than COCOM, focusing primarily on the transparency of national export control regimes. It presently has 41 members including many former COMECON countries. China and Israel, though significant arms exporters are not members but follow the groups export control regime.

It seeks to promote transparency and greater responsibility in transfers of conventional arms and dual-use goods and technologies, and to prevent destabilizing accumulations. Members seek to ensure that transfers of these items do not contribute to the development or enhancement of military capabilities which undermine these goals. Members regularly exchange information on deliveries of conventional arms to non-Wassenaar members that fall under eight broad weapons categories: battle tanks, armored combat vehicles (ACVs), large-caliber artillery, military aircraft, military helicopters, warships, missiles or missile systems, and small arms and light weapons.

India's membership in the WA has received the support of the US. It would also reinforce India's credentials in supporting non-proliferation and against WMDs. Further, it could as a major arms buyer and seller arms[6] and of dual use technology exploit the benefits of membership. However the alignments of Indian export control systems with those of the WA needs to be completed.

## Biotechnology

The rapid advances in biotechnology have made it feasible for altering natural organisms and even synthesizing genes. Such altered organisms could have features and attributes found in plant or animal species, or even new properties. The day is not far off when genetically altered organisms could produce useful products such as chemicals, bioactive substances, and fuels. However at the same time it would be possible with relatively small investments to fashion bioweapons also, for example by altering naturally occurring bacteria or viruses to make them more contagious and lethal. The prospects of such technology getting into the hands of malign non-state actors is very real. Control over such technology and materials and related equipment becomes necessary. This is the motive behind including BW related technology and equipment in the Australia Group mandate.

Controls over BW related materials and technology is complicated because of the ease of transport across national boundaries, and the fact that even very small quantities of a bioactive material can be easily multiplied using available technology. The development of highly effective sensors and accurate intelligence and monitoring of all BW related materials, equipment, and personnel may become a necessity.

## Information Technology

The development of information technology has led to incidents of cybercrime, and cyberwar, attributed to states and non-state entities. The nature of cyber threats is that they are difficult to locate conclusively the origin and thus are easily denied by the alleged perpetrators. The increasing dependence of all human activities on information systems and their relative vulnerability makes such systems easy targets for cyber attacks. There is no comparable instrument to the CWC in the domain of cyberwarfare, to regulate the behavior of state controlled entities. Cyber info warfare is another growing trend[7], using information technology to hack into target systems and using the information obtained to inflict damage. Some attempt has been made to include software technology in the WA, but this if far from adequate. What is needed is an international convention against cybercrime and an allied technology control arrangement. It remains to be seen whether this will emerge in future.

## Technology and threats from non-state actors

The march of technology has given new instruments to old non-state actors, such as terrorist entities, criminal groups, and extremists. The use of information technology has enabled such groups to mobilize support human and material to carry out their activities. The recent success of Al Qaeda and the Islamic State in recruiting home grown terrorists by remote control is a case in point. Flow of finance and WMD related crude do-it-yourself technology is also rendered easier through the internet. Controls may be needed on malicious information sent through the internet. This will be a delicate task as deciding what information should be blocked and what should not be is a difficult question. A regime for control of malicious information may need to be developed which all countries could agree to support. In this effort, there may be divergences between democratic, liberal regimes and more authoritarian or theocratic ones, which will need to be resolved. The technical requirements for monitoring information flow and detecting defined malicious information are also formidable but may be worthwhile for the sake of combating terrorism and extremism.

## Technology controls and economic development

One of the major problems with national or international technology control regimes is the impact on legitimate economic and commercial activity. The requirements imposed by controls such as licensing, etc. add a burden on firms involved with the affected products and services. At the human level, controls may affect the movement of qualified experts across national borders for high tech activities, due to cumbersome bureaucratic clearance procedures. For example the US scientific and business community has spoken out against delays in processing visas for foreign scientists coming for research projects and collaboration activities.

Economic development in developing countries has been identified as an urgent necessity. This will not only raise the living standards of a large population bust also generate demand for a wide variety of products and services which will benefit business in the developed countries and create jobs. Technology controls may sometimes be exploited by business interests in order to extract very high benefits, for example high cost of life saving drugs, which may militate against the public interest. Civil society groups have been of great help in raising the alarm and putting effective pressure. The UN has recently identified a set of goals for sustainable development by 2030, which includes access to and use of technology in critical areas. The regime for access to such critical technology must be supportive rather than obstructive and exploitative.

## Future Challenges

The rise of non-state actors and the risk of them gaining access to WMD related technology is a formidable challenge. Scenarios in which such actors gain control of nuclear materials, dangerous pathogens or chemicals, and cyber weapons are no longer in the realm of fiction. While technical means to deny access to such technology can be devised, sometimes imposing costs, the basic causes and motivations of such actors remains to be tackled and support drained away. Technology for preserving the planet from global warming and irreversible damage is urgently required to be developed and also made available freely for use on a large scale. Advances in technology will no doubt bring new challenges of regulation and control for governments and the international community.

## Endnotes

1  A Memorandum of Understanding (MoU) was signed between the US and India in 1984 on transfer of technology in exchange for alterations to India's own export-control regulations. Under this agreement, sensitive technology transfers took place such as engines for the light combat aircraft and Indian naval vessels, night vision devices for tanks etc. In the period 1984-88, there was a five-fold increase in US government approvals of civilian technology exports to India. The MoU did lead to a surge of technology licences to Indian companies and government institutions, but mainly for the items that were below the level of state-of-the-art technology. See: Indo-US Strategic Convergence: An Overview of Defence and Military Cooperation, Ashok Sharma, CLAWS, 2008, http://www.claws.in/images/publication_pdf/CLAWS%20Papers%20No[1].2,%202008.pdf ,accessed 29-3-2017

2  Between 1971 and 1974, a group of 15 nuclear supplier states held a series of informal meetings in Vienna chaired by Professor Claude Zangger of Switzerland, to reach a common understanding on equipment or material related to special fissionable material and the conditions and procedures that would govern exports of such equipment or material. The group, which became known as the Zangger Committee, decided that it would be informal and that its decisions would not be legally binding upon its members.

3  France joined the NPT in 1992 but was part of the NSG since its inception in 1975

4  The fact that Iraq claimed it had respected the NPT and IAEA agreements led to moves to go beyond what the arrangements under IAEA.

5   The US, UK, France, Russia, China, Germany and the EU

6   India's defence industry has been opened up to foreign participation. The government is also promoting a "Make in India" strategy for boosting Indian manufacturing including in defence sector. A key role would be played by dual use technology.

7   In the 2016 US Presidential elections, Russian hackers were accused of hacking into the systems of the Democratic National Committee and releasing information damaging to the Democratic Party.

# Chapter 15

# International Scientific Cooperation – Can Science bring us together ?

*"The soul of engineering science and technology lies in openness. Our time calls for peace, development and win-win cooperation. Increasing the internationalization level of engineering science and technology has become a common understanding of all countries and an important way for them to promote engineering innovation; sharing the fruits of engineering advances is vital to the common development and prosperity of all. We should strengthen international cooperation and learn from and inspire each other to promote the advancement and innovation of engineering science and technology, address common challenges of humankind and realize the common development of all nations."*

*– President Xi Jin Ping of the Peoples Republic of China, speech on 3 June 2014*

## Introduction

The last century witnessed unprecedented advances in science and technology. Scientific discoveries were earlier made by individuals or small groups of scientists in laboratories with relatively lower investments. As science advanced, the scale, complexity, and cost of scientific research expanded, resulting in increasingly larger research programmes involving many institutions and scientists. For example, nuclear science required large research infrastructure and large teams of researchers and engineers, as in the case of the Manhattan Project. In space science research and development required large scale well managed and organized efforts for which countries set up special agencies with numerous research and development centres. Such large efforts required large budgets and support of governments. In a number of cases the expenditures and technical efforts required for research programmes were such that countries found it beneficial to collaborate on such programmes rather than go ahead on their own. Even the largest economy the US found to useful to engage in such collaboration. International Scientific Collaboration took shape in several large projects, with their special features, as described in this Chapter.

## The Manhattan Project

This project was launched in order to develop nuclear weapons. It was an initiative of the US and its ally the UK and Canada, and had a clear strategic objective. Though the basic science underlying the nuclear bomb had been discovered, i.e. nuclear fission and the accompanying large release of energy, the task of developing this into a nuclear weapon required intensive efforts involving several branches of engineering, physics, chemistry, and mathematics. The project was under the direction of the U.S. Army Corps of Engineers while the nuclear physicist J. Robert Oppenheimer was the director of the Los Alamos Laboratory that designed the actual bombs. Beginning in 1939, the project grew to employ more than 130,000 people and cost nearly US $2 billion (about $27 billion in 2017 dollars). Over 90% of the cost was for building factories and producing the fissile materials, with less than 10% for development and production of the weapons. Research and production took place at more than 30 sites across the United States, the United Kingdom and Canada.

The first nuclear device ever detonated was an implosion-type bomb at the Trinity test, conducted at Alamogordo, New Mexico in July 1945. Little Boy and Fat Man bombs were used a month later in the atomic bombings of Hiroshima and Nagasaki, respectively.

In the immediate postwar years, the Manhattan Project conducted weapons testing at Bikini Atoll, developed new weapons, promoted the development of the network of national laboratories, supported medical research into radiology and laid the foundations for the nuclear navy. It maintained control over American atomic weapons research and production until the formation of the United States Atomic Energy Commission in January 1947.

The Manhattan Project was driven by the urgency of building the nuclear weapon earliest possible, despite the large cost and human effort involved. Although led by the US it involved three countries and many scientists who were earlier working in European institutions. Interestingly, among the scientists there was difference in opinion as to whether such a weapon once developed should actually be used causing huge civilian deaths and injuries. The decision to use the weapon twice against Japan remains controversial.

## CERN

In the field of particle physics, progress required the building of more and more powerful particle accelerators capable of producing particle collisions at higher and higher energies to reveal the mysteries of the fundamental nature of matter and forces, the "new physics". For a while the US led the race with the world's most powerful accelerator, the Tevatron at Fermilab, at Batavia, near Chicago[1] until a more powerful accelerator the Large Hadron Collider (LHC)[2] built by the European Organization for Nuclear Research (CERN) started operating in 2010. An even more powerful accelerator the Superconducting Super Collider (SSC)[3] was planned to be built by the US in Texas. The project was launched in 1983, but the US Congress officially canceled it in October 1993 after $2 billion had been spent due to rising cost estimates (from $ 4 billion to $12 billion), and other factors. This illustrates the fact that no single country could afford the very high costs of such research facilities.

CERN, established in 1954 as an intergovernmental agency, near Geneva on the Franco–Swiss border, now has 22 member states.

Israel is the only non-European country granted full membership. In 2016[4] CERN had 2,531 staff members, and hosted some 13,128 fellows, associates, apprentices as well as visiting scientists and engineers representing 608 universities and research facilities. Its budget in 2016 was CHF 1.1 billion funded by contributions from its 22 members, ranging from 20.6 percent from Germany (largest contributor) to 0.3 percent from Bulgaria (the smallest contributor). The Council, on which all member states are represented, is the top decision making body of CERN.

CERN's main function is to provide the particle accelerators and other infrastructure needed for high-energy physics research – as a result, numerous experiments have been constructed at CERN through international collaborations. CERN is also the birthplace of the World Wide Web. It has a large computer facility containing powerful data processing facilities, primarily for sharing and processing huge volumes of data from its experiments with researchers elsewhere through a large network. Since its foundation by 12 members in 1954, CERN's membership has expanded. Spain which joined CERN in 1961, withdrew in 1969, but rejoined in 1983. Yugoslavia, a founding member withdrew in 1961. Of the 22 members, Israel joined CERN as a full member in 2014, becoming the first non-European full member. Cyprus, Serbia, Turkey, Ukraine, Pakistan and India are associate members, while Japan, Russia and the US are observers. In addition CERN has cooperation agreements with a large number of non-member countries.

CERN operates several major facilities, which are active in the field of high energy and particle physics. It also has a large group of theoretical scientists, and exchanges data worldwide with its collaborating institutions through its computer network. In fact CERN was the first to develop the web for this purpose. CERN has been credited with a number of discoveries such as the W and Z bosons (1983), for which the 1984 Nobel Prize for Physics was awarded, and the Higgs boson (2012). Among the spin offs of CERN's research are licensed technologies[5] for computing, cryogenic techniques, detectors and sensors, electronics and chips, etc. The generation of useful new technology is an important factor in justifying the large public expenditure on CERN.

CERN has been a successful in terms of fundamental research. It has brought together a formidable network of scientists across the world, and enabled developing countries to benefit from its work through training, research exposure, and project participation. India has been collaborating with CERN since the 1960s and has recently become an associate member of CERN in November 2016. India has participated in the construction of the Large Hadron Collider (LHC), in the areas of design, development and supply of hardware accelerator components/systems and its commissioning and software development and deployment in the machine. Indian scientists have played a significant role in one of the two large experiments that led to the discovery of the Higgs Boson. The collaboration arrangement is mutually beneficial. India gets credit at European prices for its supply of equipment, which can then be used to finance the expenses of Indian researchers at CERN. India's associate membership of CERN will enable it to play a more active role in it.

## International Space Station (ISS)

The International Space Station (ISS) is a space station or a habitable artificial satellite, in low Earth orbit (400 km height, 93 minutes per orbit). Its first component was launched into orbit in 1998, and the ISS is now the largest artificial body in orbit and can often be seen with the naked eye from Earth. The ISS has a mass of 420 kilogrammes, and consists of pressurized modules, external trusses, solar arrays, and other components. ISS components have been launched by Russian Proton and Soyuz rockets, and American Space Shuttles.

Five different space agencies representing 15 countries built the $100-billion International Space Station[6] and continue to operate it today. USA's NASA, Russia's Roscosmos State Corporation for Space Activities (Roscosmos), the European Space Agency, the Canadian

Space Agency and the Japan Aerospace Exploration Agency are the primary space agency partners on the project. The ownership and use of the space station is established by intergovernmental treaties and agreements. The station is divided into two sections, the Russian Orbital Segment (ROS) and the United States Orbital Segment (USOS), which is shared by many nations. ISS is presently being funded until 2020. Discussions to extend the space station's lifetime are ongoing among all international partners; several countries, such as Canada, Russia and Japan, have expressed their support for extending the station's operations.

The ISS serves as a microgravity and space environment research laboratory in which crew members conduct experiments in biology, human biology, physics, astronomy, meteorology, and other fields. The station is suited for the testing of spacecraft systems and equipment required for missions to the Moon and Mars.

The ISS is the ninth space station to be inhabited by crews, following the Soviet and later Russian Salyut, Almaz, and Mir stations as well as Skylab from the US. The station has been continuously occupied for 16 years plus since the arrival of Expedition 1 on 2 November 2000. The station is serviced by a variety of visiting spacecraft: the Russian Soyuz and Progress, the American Dragon and Cygnus, the Japanese H-II Transfer Vehicle, and formerly the Space Shuttle and the European Automated Transfer Vehicle. It has been visited by astronauts, cosmonauts and space tourists from 17 different nations.

The legal structure that regulates the station is multi-layered. The primary layer establishing obligations and rights between the ISS partners is the Space Station Intergovernmental Agreement (IGA), an international treaty signed in 1998 by fifteen governments involved in the Space Station project. The ISS consists of Canada, Japan, the Russian Federation, the United States, and eleven Member States of the European Space Agency (Belgium, Denmark, France, Germany, Italy, The Netherlands, Norway, Spain, Sweden, Switzerland and the United Kingdom). The second layer of agreements between the partners are referred to as 'Memoranda of Understanding' (MOUs), of which four exist between NASA and each of the four other partners. There are no MOUs between ESA, Roskosmos, CSA and JAXA because NASA is the designated manager of the ISS. The MOUs are used to describe the roles and responsibilities of the partners in more detail. A third layer consists of bartered contractual agreements or the trading of the partners' rights and duties, including the 2005 commercial framework agreement between NASA and Roskosmos that sets forth the terms and conditions under which NASA purchases seats on Soyuz crew transporters and cargo capacity on unmanned Progress transporters. A fourth legal layer of agreements implements and supplements the four MOUs further. Notably among them is the ISS code of conduct, setting out criminal jurisdiction, anti-harassment and certain other behavior rules for ISS crewmembers, made in 1998.

The ISS project could be an example for future extra terrestrial establishments, based on international cooperation. Both the US and the Soviet Union found it better to cooperate in the high cost effort to build a habitat in space, and this cooperation increased with the end of the Cold war. Other countries joined the effort, contributing to its success as a fine example of sharing of knowledge and research. Programmes such as the ISS may also help to bridge differences between countries, for example between the US and Russia. However there is uncertainty over the future of the ISS programme.

Could India join the ISS? South Korean and Indian space agencies had expressed some interest in participation the ISS program in 2009. India has a major space programme led by ISRO and collaborates already with ISS partners. There has been some speculation on this subject. However ISRO has given priority to unmanned space missions which are much lower cost and for joining the ISS, ISRO will need substantial budget increase. Though the ISS affords considerable savings for manned space exploration, over a separate manned space programme, it still is much higher than cost of unmanned space missions. The head of the European Space

Agency has said in an interview that the ISS should be opened up to astronauts from India and China. Future increase of the number of partner space agencies in the ISS would make it more economically viable.

## International Thermonuclear Experimental Reactor (ITER)

ITER, located in France, is an international nuclear fusion research and engineering megaproject, which will be the world's largest magnetic confinement plasma physics experiment. It aims to develop full-scale electricity-producing fusion power stations. Fusion has been accomplished in the case of the thermonuclear bomb with enormous destructive capacity, but the challenge is to achieve controlled fusion and produce energy for useful purposes. ITER seeks to achieve this by confining a plasma (extremely hot ionized gas) within a magnetic field long enough to produce fusion energy. The ITER fusion reactor[7] has been designed to produce 500 megawatts of output power for at least 1000 seconds while needing 50 megawatts to operate and thus demonstrate the principle of producing net energy from the fusion process. If successful it could lead to fusion as a large-scale and carbon-free source of energy based on the same principle that powers the Sun and the stars.

The project was conceptualized in 1985 and is funded and run by seven member entities—the European Union, India, Japan, China, Russia, South Korea, and the United States. The EU, as host party for the ITER complex, is contributing 45.6 percent of the cost, with the other six parties contributing approximately 9.1 percent each, with 90 percent of the contribution being delivered to the ITER Organization in the form of completed components, systems or buildings.

Construction of the ITER Tokamak complex started in 2013 A project review in 2015 revealed time overruns of six years and cost overruns and performance of the project management has been criticized. As of 2016, the total price of constructing ITER is expected to be in excess of Euro 20 billion compared to initial estimates of the proposed costs for ITER of Euro 5 billion for the construction and Euro 5 billion for maintenance and the research connected with it during its 35-year lifetime. Although Japan's financial contribution as a non-hosting member is 9.1 percent of the total, the EU agreed to grant it a special status, under the Broader Approach Agreement of 2007[8], under which Japan will provide for 18 percent of the research staff at Cadarache and be awarded 18 percent of the construction contracts, while the European Union's staff and construction components contributions will be cut by the same extent. This was a compromise in order to reach agreement on the site of ITER in the EU, for which Japan had also bid. The U.S. withdrew from the ITER consortium during 1998-2003, mainly due to concerns over time and cost overruns, poor project management, and feasibility of the project in the Congress. Following management reforms, the Department of Energy has supported participation at least until 2018.

India's participation in ITER is handled by ITER-India, presently located within the premises of Institute for Plasma Research (IPR), Ahmedabad. India is supplying nine different packages, including cryostat, cooling water systems, vessel in-wall shielding blocks, radio frequency heating sources, cryodistribution and cryolines, power supplies, diagnostic neutral beam system and some of the diagnostics systems, in collaboration with Indian industry. Participation in ITER will help to accelerate India's own effort[9] to develop fusion power using a similar reactor. It will also provide opportunities to Indian industry to gain experience in high technology related areas.

ITER represents a great effort to find unlimited energy through the same process that the produces energy in the sun and stars. If successful, it could provide a way to deal with global warming while satisfying the need for energy. Given the formidable technical challenges in the ITER project international collaboration is the best way forward.

## Human Genome Project

The Human Genome Project (HGP) was an international scientific research project with the goal of determining the sequence of base pairs (about 3.3 billion) that make up human DNA, and of identifying and mapping all of the genes of the human genome from both a physical and a functional standpoint. The US government supported the concept in 1984 and the $ 2.7 billion project was formally launched in 1990 and was declared complete in 2003, two years ahead of schedule. Funding came from the US government as well as numerous other groups from around the world. A parallel project was conducted outside of government by the Celera Corporation, or Celera Genomics, which was formally launched in 1998. Most of the government-sponsored sequencing was performed in twenty universities and research centers in the United States, the United Kingdom, Japan, France, Germany, Canada, and China.

The sequencing of the human genome promises benefits in many fields, from molecular medicine to human evolution. The HGP, through its sequencing of the DNA, can help understand diseases including, genotyping of specific viruses to direct appropriate treatment; identification of mutations linked to different forms of cancer; the design of medication and more accurate prediction of their effects; advancement in forensic applied sciences; befouls and other energy applications; agriculture, animal husbandry, misprocessing; risk assessment; bioarcheology, anthropology and evolution. Another proposed benefit is the commercial development of genomics research to develop DNA based products, a multibillion-dollar industry. Since the HGP, the gene sequencing technology has evolved rapidly[10] and presently the genome of an individual can be sequenced at a cost of below $1500.

## SESAME Project

An interesting initiative is the Synchrotron-Light for Experimental Science and Applications for the Middle East (SESAME) project, a "third-generation" synchrotron light source under construction in Allan near Amman, Jordan. It is the Middle East's first major international scientific research centre as a cooperative venture by scientists and governments of the region set up on the model of CERN (European Organization for Nuclear Research) promoted by UNESCO (United Nations Educational, Scientific and Cultural Organization). It is an autonomous intergovernmental organization.

SESAME will (1) Foster scientific and technological excellence in the Middle East and neighbouring countries in subjects ranging from biology, archaeology and medical sciences through basic properties of materials science, physics, chemistry, and life sciences; and (2) Build scientific and cultural bridges between diverse societies, and contribute to a culture of peace through international cooperation in science. The main facility is a 133 metre diameter particle accelerator of up to 2.5 GeV energy, which will produce intense beams of radiation covering a wide range from infra red to X-rays ( up to 100 KeV), making it a highly versatile facility with applications in many branches of science. Despite financial, technical infrastructural obstacles, political problems in the region and delays, the project is likely to be inaugurated in May 2017.

The current (2015) Members of SESAME are Bahrain, Cyprus, Egypt, Iran, Israel, Jordan, Pakistan, the Palestinian Authority, and Turkey. Current Observers (2015) are Brazil, China (People's Republic of), the European Union, France, Germany, Greece, Italy, Japan, Kuwait, Portugal, Russian Federation, Spain, Sweden, Switzerland, the United Kingdom, and the United States of America. The capital cost of some $110 million of the project as well as the eventual annual operating cost of $ 6 million would be met through voluntary contributions from member countries and other donors. An important objective behind the project is to stimulate cooperation and understanding among the member countries, which includes Israel and several Arab states. Whether it will succeed in this objective remains to be seen, but it has been regarded as important for the advancement of scientific research in the region.

## Science, Technology and development issues

There is no doubt that science and technology advances have greatly increased international trade and business and contributed to economic development. The share of manufactured products in world trade has increased[11] from 58 per cent (1965) to 77 per cent (2004). Technology advances, combined with globalization of technology, liberalization of trade in goods and services, have led to growth of trade in high tech products and services, and the rise of global supply chains. There is a demand for increasingly freer movement of highly skilled labour to complement the mobility of capital. The pace of change has been increasing posing challenges to businesses and governments. In the 21 st century, growth will be increasingly driven by investment in knowledge-based capital (KBC)[12]. These intangible assets include digital information (software and data), innovative property (patents, copyrights, trademarks and designs) and organisation-specific competencies (brand equity, training and organisational capital). They are a key factor to create the types of innovation that spur new sources of growth.

This is despite the controls and restrictive licensing practices attached to technology transfers to partners in the developing world. Technological advances by large corporations have often been used to gain commercial advantages and profits. There is opposition to international efforts to regulate the conditions for transfer of technology and make it more in line with the needs of economic development and reduction of inequality. For example, the US and OECD countries opposed the efforts of the G-77 group of developing countries under UNCTAD to evolve a binding code of conduct for transfer of technology[13] which also sought to regulate restrictive business practices analogous to domestic legislation on competition policy. The argument was that technology was largely in the hands of private business and transfer arrangements were best left to the business partners concerned. However this ignored the great difference in bargaining power of the seller and the recipient of technology, and the unfair restrictions (which would not be permitted under domestic competition policy) that could be imposed on the latter.

Can the international cooperation framework in science and technology be adapted to enable reduction of inequality among and within countries and help solve the urgent development issues whcih have been recently expressed in terms of the UN's Sustainable Development Goals 2030 ? To meet this challenge, private sector and public agencies that have relevant technology should be encouraged to share development related technology on fair and equitable terms. For example, given the urgent need to reduce carbon emissions, it is essential that low carbon technology be made available across the world at the lowest possible cost. Another example is the issue of access to affordable life saving drugs, medicines, and health care services to needy populations across the globe.

There is an effort address such issues by evolving technology sharing platforms[14] relevant to the SDG 2030, but much more needs to be done. It should be recognized that in the struggle to achieve the SDG 2030, alternative perspectives based on global cooperative efforts rather than competition for economic advantages is necessary. The free sharing and dissemination of technology relevant to SDG 2030 would expand the pool of human knowledge and benefit all. Growth in affordable technology for social and global good could help boost economic growth. Global problems require a cooperative effort where sharing of technology is seen as leading to expanding the pool of benefits to be shared, in contrast to a competitive approach where the possessors of technology seek to extract maximum short term benefits from it. There is an urgent need for greater efforts in research and innovation focused on more inclusive development that meets the needs of lower income populations, and build strategic alliances and partnerships to achieve this objective.

## Future outlook

The research required to expand the frontiers of knowledge is becoming increasingly demanding in terms of financial and human resources.

Therefore in future, one can expect such cutting edge research will be done by international consortia comprising institutions of many countries, with government as well as private funding, and sharing of the fruits of such research. Could this bring about a new era of peace and cooperation among countries? Certainly scientific communities across the world could push governments in this direction. This would be the spirit of renewed technodiplomacy, where technological cooperation stimulates better diplomatic relations.

In South Asia, could it be possible to bridge the differences between countries such as India and Pakistan through scientific collaboration? The idea is certainly worth exploring, since there are important areas of technology where both countries could have an interest. Some promising areas might be renewable energy, marine ecosystem management, human health, and ICT applications for development, environment protection, climate change, etc. The potential for scientific collaboration in South Asia has not been expoited so far, and perhaps the future may witness more progress.

## Endnotes

1  The Tevatron accelerated protons and antiprotons in a 6.86 km, ring to energies of up to 1 TeV (Tera electron volts or 1000 billion electron volts of energy), hence its name. The Tevatron was completed in 1983 at a cost of $120 million and significant upgrade investments were made in 1983–2011. It was closed down in 2011, after having contributed to several major discoveries in particle physics. The Tevatron ceased operations in September 2011, due to budget cuts and because of the completion of the LHC, which began operations in early 2010 and is far more powerful (planned energies were two 7 TeV beams at the LHC compared to 1 TeV at the Tevatron).

2  The LHC was built between 1998 and 2008 in collaboration with over 10,000 scientists and engineers from over 100 countries, as well as hundreds of universities and laboratories. It lies in a tunnel 27

kilometres in circumference, beneath the France–Switzerland border near Geneva, Switzerland. Its initial energy was 3.5 teraelectronvolts (TeV) per colliding beam (7 TeV total), almost 4 times more than the previous world record for a collider, rising to 4 TeV per beam (8 TeV total) from 2012. The LHC was upgraded and restarted in April 2015, reaching 6.5 TeV per beam on 20 May 2015 (13 TeV total, the current world record).

3 The Superconducting Super Collider (SSC) was a particle accelerator complex under construction in the vicinity of Waxahachie, Texas. Its ring circumference of 87.1 kilometers and an energy of 20 TeV per proton would have made it the world's most powerful, surpassing by far the Large Hadron Collider. The project was cancelled in 1993 due to budget problems.

4 CERN: Quick Facts 2016, https://cds.cern.ch/record/2152342/files/CERN-Brochure-2016-002-Eng.pdf , accessed 15-1-2017

5 CERN has an active platform for licencing technologies and knowledge transfer. See Technology Portfolio, CERN Knowledge Transfer website, http://kt.cern/technology-transfer/external-partners/portfolio , accessed 16-1-2017

6 International Space Station: Facts, History & Tracking, space.com, Tim Sharp, 5 April 2016, http://www.space.com/16748-international-space-station.html , accessed 16-1-2017

7 What is ITER, ITER website, https://www.iter.org/proj/inafewlines , accessed 16-1-2017

8 The Broader Approach Agreement of 2007 between Euratom and Japan covers research and technology activities supporting three large projects in Japan that are related to the ITER project . The Broader Approach, European Commission, http://ec.europa.eu/research/energy/euratom/index_en.cfm?pg=fusion&section=broader , accessed 28-3-2017

9   Indian Fusion Programme and Contribution for ITER Project, A Mukherjee, ITER-India, Octpber 2015, http://basharesearch.com/WCSET2015/wcset2015009.pdf , accessed 17-1-2017

10  The Cost of Sequencing a Human Genome, National Human Genome Research Institute, https://www.genome.gov/sequencingcosts , accessed 17-1-2017

11  Technology, globalization, and international competitiveness: Challenges for developing countries, Carl Dahlman, Industrial Development in the 21 st century, Sustainable Development KNowledge Platform, 2006, http://www.un.org/esa/sustdev/publications/industrial_development /1_2.pdf , accessed 19-4-2017

12  Science, Technology and Industry, OECD, 2014, https://www.oecd.org/sti/sti-brochure.pdf, accessed 19-4-2017

13  Negotiations on a draft International Code of Conduct on the Transfer of Technology (ToT) took place under UNCTAD auspices between 1976 and 1985. The draft was not adopted, but its principles influenced how ToT is reflected in international law and policy today, specifically in Article 66.2 of TRIPS, Convention on Biodiversity and Nagoya Protocol benefits, and the WHO's Global Strategy and Plan of Action on Public Health, Innovation and Intellectual Property.

14  Technology Facilitaiton Mechanism, UN Sustainable Development Knowledge Platform, https://sustainabledevelopment.un.org/TFM/STIForum2017 , accessed 19-4-2017

# Chapter 16

# Managing Science and Technology Diplomacy

*"Japan upholds the following four basic concepts for the strategic promotion of science and technology diplomacy: (1) bilateral and multilateral cooperation to promote science and technology and innovation, (2) utilization of science and technology for solving global challenges, (3) promotion of bilateral relations through science and technology cooperation, and (4) promotion of "soft power" as a science and technology-oriented country."*

*— Ministry of Foreign Affairs Japan, on Science and Technology Diplomacy*

## Introduction

As science and technology advances at an increasing rate, and becomes more and more important in national development, growth, and competitiveness, there is likely to be a corresponding increase in interactions between countries in this field. Engagement with other countries could have various objectives – joint research activities, gaining access to science and technology, and licensing and sale of technology, to cite a few examples. Just as trade and investment have an important influence in relations between countries, science and technology exchanges can also have an impact. Therefore it becomes important how a country structures and manages its external engagement in the domain of science and technology so as to meet its overall foreign and national policy objectives, and exploits the opportunities offered. In many cases the architecture of science and technology engagement has developed without long term planning, driven largely by events and immediate needs. The countries advanced in science and technology, such as the US, Japan, Europe, etc has developed specific architecture to meet their objectives. India also has to develop its systems for managing engagement in science and technology with both the advanced countries as well as the developing countries, in order to meet its aspirations and objectives.

## Evolution of Modern Diplomacy

Modern diplomacy has evolved considerably from the time when diplomats dealt only with political matters and only with foreign ministries. Today's diplomacy has to deal with a host of subjects, such as consular affairs, economic and business affairs, media and information, education, culture, security and defence, and increasingly science and technology. Since the staff strength available is limited diplomatic missions often have personnel handling several subjects, backed by the concerned agencies back home. Effective functioning therefore requires good coordination with other concerned government and nongovernmental agencies together with some degree of subject knowledge with the personnel in the field. The advent of the internet and electronic communications has greatly facilitated this especially for diplomatic missions of developing countries. While operations have become easier, however, strategic planning and identification of long term objectives in various subject areas in the engagement with a country remains a challenge. In many cases such engagement is largely driven by events such as high level visits, resulting in a lack of coherence and less than optimum benefits.

## Science and Technology in Foreign Relations

Many areas of science and technology have become prominent in international relations, and this has been covered in detail in the earlier chapters. Technology can determine a country's military capability, as well as its economic strength and competitiveness. It can be disruptive, causing large scale job changes and start a new cycle of winners and losers. Very often advances in technology start among the scientists concerned, then spread to the business environment, and impinges on the consciousness of policy makers and the public only when some dramatic effects emerge. The foreign policy responses to such changes are often ad hoc and improvised, driven by short term considerations. An example is climate change, where the lack of immediate dramatic effects leads to complacency in tackling a serious global challenge. Scientists and policy makers need to have better dialogue on long term consequences of new technological developments, including on international relations.

'Science diplomacy' is still a fluid and evolving concept, but can usefully be applied to the role of science, technology and innovation in three dimensions[1] - (1) Scientific advice and inputs into foreign policy making (science in diplomacy); (2) Promoting international science cooperation (diplomacy for science); and (3) Using science cooperation to improve relations between countries (science for diplomacy). To this one may add a fourth important dimension - using science and technology cooperation for sustainable development (science for sustainable development).

## Scientific Diplomacy Management

In most countries the government system for science and technology consists of a Ministry handling this subject, together with a number of government funded research institutions and universities. The Ministry of S & T often has a department for international S & T cooperation, which handles external engagement in this sector. Most of this engagement is through visits of delegations, conferences and seminars, etc., and some of these contacts may result in more intensive engagement such as joint projects. Framework bilateral agreements for cooperation in S & T are generally concluded, under which programmes of cooperation covering specific activities to be carried out by specified partner institutions are negotiated, for multiyear durations and revised periodically. Generally the financing of activities is shared by partners on both sides, as also the benefits emerging from cooperation activities. The cooperation activities may be seminars, workshops, exchanges of experts, training, research and development. The IPR benefits emerging from research are shared by both sides. Such models of bilateral S & T cooperation are commonly worked out between governments of both sides with the Ministries concerned with S & T playing the leading role.

At the field level, governments may have science attaches positioned in their Embassies located in countries where there is significant interaction. A science attaché has a role similar to attaches dealing with subjects such as commerce, defence, culture, etc. and works under the overall supervision of the Head of Mission. The functions of a science attaché depend on the conditions in the sending country and the receiving country, but broadly are (a) collecting, analyzing and reporting information on S & T in the host country; (b) promoting and protecting the national interests of the sending country as regards the S & T sector and (c) providing advice on S & T related issues for foreign policy formulation and implementation. The US State Department's role in S & T was covered in a1950 report[2] in some detail, with an emphasis on intelligence collection related to S & T, which was corrected in later years. In practical terms the science attaché was expected to advise the chief of mission on scientific and technical matters, report in accordance with instructions, and represent the chief of mission and the government in scientific and related affairs. This leaves considerable room for interpretation and confusion. Finding staff to man these posts poses problems. Highly qualified scientists are too narrowly focused and may not be willing to detach themselves from their research activities for long periods. Also scientists posted in countries may face problems with the local language and adapting to living in strange countries and diplomatic practice. At the same time the internet has provided real time access to data that makes it possible to collect much S & T related information from public sources. Therefore the role of the scientific attaché and related management issues need to be continuously reviewed and adjusted.

S & T activity takes place within an ecosystem where support institutions play a great role in supporting and commercializing S & T research and innovation. Today the early acquisition of rights to promising S & T research and innovation can confer valuable advantages. Therefore it becomes crucial to scan the S & T sector and provide early information on promising breakthroughs that may have commercial or strategic value, and to alert potential partners back home. This role of the science attaché has been given greater importance by many countries. Basically the business and strategic opportunity dimension of S &T research and innovation needs to be captured in time and ahead of potential competitors and exploited. Therefore the S & T attaché needs to be aware of ongoing developments within the host country's S&T ecosystem as a whole, as well as being aware of the state of development of the home country S & T and business ecosystem. This involves interacting with research institutions, supporting agencies, public as well as private, on a continuous basis. A highly qualified scientist may find it difficult to adapt to such a role. On the other hand, a diplomat with a background in S & T may find it easier to adapt to and handle such a role, given some training and support. A background in S & T may be an asset as it trains an individual to rapidly collect and analyze information.

Home country support in the form of updates and changes in the requirements of the home country S & T ecosystem and related business ecosystem would need to be provided. Ideally a public-private partnership entity could provide such backing from the home country. Such an entity could be an important functional interface for the day to day work of the science attaches abroad, analogous to the Chambers of business and the commercial attaches.

## Country Examples

A detailed examination of the S & T sector in countries and its diplomatic projection is beyond the scope of this book. However a few aspects can be highlighted. The situation in Russia and Israel may illustrate the differing approaches being followed.

The US being the most advanced in S & T launched a scientific dimension to its diplomacy in the 1950s. The State Department was given the mandate to run a programme of science attaches, supported by a Science Office in headquarters, and to work together with the CIA and other institutions such as the National Academy of Sciences. The focus on intelligence collection and the negative perceptions it generated led to changes in later years. The programme faced budget constraints, and by 1955, only 4 science attaches were placed in London, Stockholm, Paris, and Tokyo. The launch of the Soviet satellite Sputnik in 1957 spurred action and the number of science attaches grew to 23 in 1966 located in 17 Missions, with 6 having two attaches each. However headquarters support in the form of a Science Officer proved difficult to implement, and staffing and management problems remained. After the end of the Cold War, the programme was reduced and finally eliminated in 1997[3], and a new structure was proposed.

While the growing importance of S & T in foreign policy has been recognized, the structural and programmatic response is still evolving. The Director of the White House Office of Science and Technology Policy (OSTP)[4] provides advice the President on international S&T cooperation policies and the role of S&T considerations in foreign relations. Department of State sets the overall policy direction for U.S. international S&T diplomacy, and works with other federal agencies as needed. OSTP also acts as an interagency liaison. A number of federal agencies that both sponsor research and use S&T in developing policy are involved in international S&T policy. Within the State Department, the Bureau of Oceans and International Environmental and Scientific Affairs (OES) coordinates international S&T activities, and the Science and

Technology Advisor (STAS) provides S&T advice to the Secretary of State, Department of State staff, and the director of USAID. The goals of the STAS are to enhance the S&T literacy and capacity of DOS; build partnerships with the outside S&T community, within the U.S. government, with S&T partners abroad, and with foreign embassies in the United States; provide accurate S&T advice to DOS; and shape a global perspective on the emerging and "at the horizon" S&T developments anticipated to affect current and future U.S. foreign policy.

In June 2009, President Obama announced several international S&T diplomacy programs in Muslim-majority countries including a new fund for technological development in these countries, establishing centers of scientific excellence, and appointing new science envoys. This was an example of a science for diplomacy outreach to the Muslim countries. The same month, the US House of Representatives passed the International Science and Technology Cooperation Act of 2009 which would require the OSTP Director to establish an interagency committee to identify and coordinate international science and technology cooperation.

The responsibilities of scientific diplomats vary widely, depending in large part on the relative positions of the two nations and their level of scientific interchange. A US science officer in Paris, for example, typically spends much time dealing with cooperative space programs and other bilateral agreements. In a country such as Egypt, a science attaché from a Western embassy would be likely to monitor development cooperation projects. The Federal Republic of Germany posts four science counselors to Washington. One is an engineer, one a scientist and two are diplomats who specialize in the scientific side of foreign policy. France, Canada and Italy generally rely on scientists, engineers, university professors or persons from the private sector. Japan and the Soviet Union rely mostly on career diplomats who have some science expertise. Italy has located a science attache in its mission in South Africa as an attempt to build research cooperation with South Africa as well as other Sub Saharan countries in the region.

The UK had 27 science officers posted in 14 of its overseas missions (in 1984)[5], located in 11 countries. The Government Office for Science is part of the British government which advises the UK Government on policy and decision-making. It is led by the Government Chief Scientific Adviser (GCSA), who reports to the Prime Minister and Cabinet. France has one of the most comprehensive S & T diplomacy establishments. For example, the French Office for Science and Technology (OST) in the US, consists of a team of 24 staff members including professors, senior researchers and engineers located in the Embassy (Washington, DC) and 6 consular offices (Atlanta - Boston - Chicago - Houston - Los Angeles - San Francisco). In India, France has a Science and Technology department in its Embassy[6], composed of a team of eight, including two Attachés for Science and Technology respectively based in Bangalore and Mumbai. Several leading French research institutions are also represented in India - CNRS in New Delhi, CNES in Bangalore; and CEA, in New Delhi.

Denmark has set up Innovation Centers in seven cities (Munich, Seoul, New Delhi, Sao Paulo, Shanghai, Tel Aviv, and Silicon Valley) to build bridges between research institutions, companies and capital in Denmark and abroad; accelerate the entry of Danish companies into foreign markets; promote foreign investments in Denmark; and facilitate research cooperation and provide inspiration to help drive innovation in Denmark.

## Russian Federation

Russia has undergone a transition from a centrally planned and controlled economy and this has impacted its S & T sector. Russian S & T establishments were government funded and run and concentrated on strategic S & T research which enabled Russia to make formidable advances in nuclear technology, aerospace, materials, and basic sciences. After the break up of the Soviet Union, the strategic industries were dispersed among the different constituents, causing disruptions. Russia has lost some of its scientific and technological human resources, and research has suffered due to disruptions.

Russia has undertaken a major reform its S & T structures to make them more in tune with the needs of the Russian economy as it moves ahead to a more knowledge and market based economy in future. Funding does not appear to be a problem, but the management and allocation of funds for projects is a challenge. The Russian government is trying to use its funding strategy to strike a balance between purely academic research in which Russia had considerable strength, and applied research which included innovation, technology transfer and commercialization. University based research with commercialization and upgrading of universities is being promoted. R&D funding was provided to institutions and universities from 4 sources- (i) Federal Targeted Programme for 2014-2020; (ii) Russian 5-100-2020 programme to upgrade Universities; (iii) Russian Science Foundation; and (iv) Russian Foundation for Basic Research. There is interest in collaborating with India which is seen as growing rapidly with a high number of talented young students and researchers. The traditional mechanisms for cooperation between India and Russian institutions such as the Russian Academy of Sciences may need to be reformed in tune with recent developments.

## Israel

Israel presents a remarkably dynamic scientific and technical sector, despite is small size, and low population of 8 million. The requirement of mandatory military service (3 years after high school) imposes a drain on science and technology education and development of the youth, though this has helped Israel to advance in cyber security and defence systems. Of the population some 1 million are orthodox Jews and another 1 million are Arabs, both groups are relatively backward in participating in modern scientific and technological activities. Taxation in Israel is relatively high, due to the high burden of security. There is no doubt that a comprehensive peace in the Middle East would result in great dividends for Israel in terms of human and economic resources for science and technology.

An interesting fact is that Chaim Weizmann who discovered a process for making acetone from starch licenced his badly needed technology to the British during World War I, and in exchange for this he got the British to issue the Balfour Declaration in 1917, an interesting connection between science and politics. Weizmann later became the first President of Israel and played a big role in Israel's science and technology development.

Seven Israeli universities and network of research centres have resulted in an ecosystem that provides rich opportunities for startups and has attracted participation of most of the world's large corporations. The Weizmann Institute of Science is an important Israeli research university. It has a special Technology Transfer (TT) unit which develops research outputs to the point of patentability. A unit called Yadev (knowledge) scouts for promising research work and also advises research teams of gaps they should fill, an important intervention at the pre-commercialization stage. The patent benefits are shared 60-40 between the Institute and the research team. The patent development is undertaken by the TT unit, with the help of outsourced legal services. About 90 patent proposals are made each year of which some 60 go on to final patent applications.

While the Israeli market is small, Israeli startups have been bought out by leading multinationals scouting for promising startups, and making large profits through such acquisitions of technology and developing and exploiting their global market potential. Over 300 research centres have been set up in Israel by multinational corporations in various fields illustrating the good potential for R & D. In the strategic sector, nuclear, aerospace and defence industries have been Israeli strengths, generating considerable export interest. There is much to be gained by merging Israeli practices on innovation and science with India's large skilled human resource base. In this context, strengthening cooperation and exchanges between Indian

Universities and Israeli Universities could be of great benefit to students, researchers and faculty.

The Office of the Chief Scientist (OCS), located in Jerusalem in the Ministry of Science, Technology and Space (MOST)is the Science and Technology promotion agency of the Ministry. It works with the Department of Science and Technology of India,. India has a relatively small diplomatic mission in Israel, and there could be benefits in locating a full time science officer in Israel, or covering it from by a science officer based in an Indian mission in Europe. This illustrates the need to have an appropriate network of science officers for engaging with key countries advanced in science and technology.

## Science and technology in Indian diplomacy

The Office of the Principal Scientific Adviser to the Government of India (PSA's Office) was set-up in November, 1999 in order to : (1) Evolve polices, strategies and missions for the generation of innovations and support systems for multiple applications, (2) Generate science and technology tasks in critical infrastructure, economic and social sectors in partnership with Government departments, institutions and industry, and (3) Function as the Secretariat to the Scientific Advisory Committee to the Cabinet, with the Principal Scientific Adviser to the Government of India as its Chairman. The PSA networks with his counterparts in other countries providing a useful mechanism for policy level dialogue in science and technology.

The Department of Science & Technology has established an International Cooperation Division[7] with responsibility of (1) negotiating, concluding and implementing S&T Agreements between India and other countries; (2) providing interventions on S&T aspects in international forums. The Division works in close consultation with the Ministry of External Affairs, Indian Missions Abroad, and S&T Counselors at Germany, Japan, Russia and USA, stakeholders in scientific, technological & academic institutions, concerned governmental agencies and with various industry associations in India.

International Cooperation is realized through three main pillars - (1) Bilateral Cooperation with developed and developing countries, (2) Regional Cooperation such as with SAARC, ASEAN and BIMSTEC and (3) Multilateral Cooperation through EU, TWAS, IBSA, BRICS, UNESCO and NAM. Presently India has bilateral S&T cooperation agreements with 83 countries with active cooperation with 44 countries. During the recent years the cooperation has strengthened significantly with Australia, Canada, EU, France, Germany, Israel, Japan, Russia, UK and USA. Cooperation with African countries has also been strengthened through India Africa S&T Initiative. The soft power of S&T has been leveraged to engage with several countries under India's Act East policy and with some neighbouring countries.

India has set up Science Wings in Indian Missions at Berlin, Moscow, Tokyo and Washington, to provide information about the developments in research and technology and inputs arising from their meetings with researchers, government agencies and industry as well as to support the visits of scientific delegations from the respective countries. Science Wings also service the ongoing bilateral programmes of cooperation in Science and Technology and acted on behalf of various Indian agencies on matters referred to them for further action. Technical liaison officers have been posted in the Indian Missions in Austria, France, UK and USA in the fields of Space, Defence and Atomic Energy.

In the Ministry of External Affairs, the Division for Investment and Technology Promotion was set up in 1999 to include in its terms of reference, a component for addressing science and technology related issues. In the diplomatic cadre, there a substantial number of officers with S & T background whose knowledge and skills are relatively unexploited for S & T diplomacy. This is an important human resource that could be used.

At the same time, India does not have an extensive and large diplomatic network as for example the US, including specialized science officers posted abroad. Therefore, it would be logical to combine the functions of a science officer with the functions of promoting economic relations, investment, and trade. Such an officer could be designated as the officer for promotion of business, investment and technology. This would require appropriate training to be provided as well as appropriate mission guidelines from headquarters, and could provide many advantages.

## Some policy issues

There are some policy related issues in the conduct of S & T diplomacy that require clarification according to the situation of each country. Several objectives of S & T diplomacy can be identified. It may be to gain access to commercial or strategic technology for the country's national development. It may be to gain support from abroad for S & T research and development in the country through external linkages for research collaborations, attracting financing, training of researchers, and capacity building. Another objective may be to export S & T related services to other countries through various arrangements. As part of development cooperation, S & T projects may be supported in recipient countries to help their development, as well as a part of overall diplomatic strategy. Government itself may also be operating S & T institutions directly and so there may be a need to build capacity and engage in cutting edge research and development through external cooperation. Thus there are a variety of objectives which may need to be prioritized as part of an overall strategy and plan of action for each country being engaged with. These need to be regularly reviewed and updated in the light of changes.

Given the scope and scale of S & T activities, a large number of public and private institutions may be involved. Thus there is need to coordinate efforts within the overall ecosystem, firstly with government and also with the non-governmental entities that have a stake. The lead role for such coordination should be undertaken by the Ministry responsible for S & T policy. The output of such coordination needs to be translated into plans of action at the country level in the field for implementation in consultation with the Foreign Ministry and the concerned Missions. This integration of science and technology into foreign policy strategic planning is lacking.

Science provides a non-ideological environment for the participation and free exchange of ideas between people, regardless of cultural, national or religious backgrounds. Scientific values of rationality, transparency and universality can help to build trust between nations. The scientific community which often works beyond national boundaries is well placed to support emerging forms of diplomacy involving nontraditional alliances of nations, sectors and non-governmental organizations. If aligned with wider foreign policy goals, these channels of scientific exchange can contribute to coalition building and conflict resolution. Cooperation on the scientific aspects of sensitive issues can sometimes provide an effective route to other forms of political dialogue.

Science diplomacy seeks to strengthen the integrate the different interests and motivations of the scientific and foreign policy communities. Therefore it is important that scientific and diplomatic communities engage in dialogue and that goals are clearly defined. Foreign ministries should give more attention to science within their strategies, and draw more extensively on scientific advice in the formation and delivery of policy objectives. Regulatory barriers, such as visa restrictions and security controls, can also be a practical constraint to science diplomacy. Scientific organizations have an important role to play in science diplomacy, particularly when formal political relationships are weak or strained, by providing alternative channels for communication and cooperation. There need for more cooperation between policymakers, academics and researchers working in the foreign policy and scientific communities, to identify projects and processes that can further the interests of both communities. Foreign policy institutions and think tanks can play a role here.

In the field, most countries, even large countries such as the US face limitations in terms of budgets, and human resources for staffing science offices in their missions abroad. Hence there is a need to priorities regarding coverage with a limited number of offices. This may leave out many countries where there could be promising opportunities for S & T cooperation. Therefore there is a need for diplomatic officers with a science and technology background, with appropriate training and briefing, to fill this gap. Such officers can take on the functions of science officers in addition to the other responsibilities assigned to them in the mission. This would be a cost effective way of providing a network that could facilitate wide ranging S & T engagement. In either case, whether there is a dedicated science office or a part time science officer, the guidance from headquarters as to the plan of action to be followed in the S & T sector would be critical. Science officers in larger missions could be involved in working on sub regional or regional S & T activities. India's diplomatic cadre includes a substantial number of science and technology graduates, whose knowledge and skills could be tapped, enabling a more vigorous thrust to India's S & T diplomacy.

## Endnotes

1   New frontiers in science diplomacy - Navigating the changing balance of power, The Royal Society, January 2010, https://royalsociety.org/~/media/ Royal_Society_Content/policy/publications/2010/4294969468.pdf   , accessed 28-4-2017

2   The science attaché program, CIA, 18 Sept 1995, https://www.cia.gov/ library/center-for-the-study-of-intelligence/kent-csi/vol10no2/html/ v10i2a02p_0001.htm , accessed 26-1-2017

3   Diplomacy for Science Two Generations Later, Science and Diplomacy, AAAS, Igor Linkov et al, 13 March 2014, http://www.sciencediplomacy. org/perspective/2014/diplomacy-for-science-two-generations-later    , accessed 26-1-2017

4   Science, Technology, and American Diplomacy: Background and Issues for Congress, Stine, D.D., Congressional Research Service report RL35403, 29 June 2009, https://fas.org/sgp/crs/misc/RL34503.pdf , accessed 19-4-2017

5   House of Commons, Written answers, 4 April 1984,  http://hansard. millbanksystems.com/written_answers/1984/apr/04/scientific-attaches , accessed 26-1-2017

6   S & T Department, Embassy of France in India, http://www.ambafrance-in.org/Science-and-Technology-Department , accessed 26-1-2017

7   International S & T cooperation, Department of Science and Technology, Government of India, http://www.dst.gov.in/international-st-cooperation , accessed 26-1-2017

# Chapter 17

# Conclusion – Future Outlook and Challenges

*"Now, our planet is facing several global challenges: to its atmosphere, to its resources, to its inhabitants. Wicked problems such as climate change, over-population, disease, and food, water , energy and cyber security require worldwide collaboration to find sustainable solutions. It is science that provides our understanding of these issues, and it is science that will underpin our solutions. But as the climate change 'debate' demonstrates, these are no longer solely scientific and technical matters. Solutions must be viable in the larger context of the global economy, global unrest and global inequality. In short, the solutions need to be based not only on sound science, but on sound politics as well. It stands to reason, then, that scientific expertise should be a fundamental part of diplomatic efforts. As single nations can neither solve them alone nor develop solutions to every problem, scientific cooperation becomes an increasing necessity."*

*– Dr Alan Finkel, Chief Scientist of Australia, 2016*

## Introduction

Scientific and Technological development has grown at an increasingly rapid pace especially since the dawn of the 20 th century. This trend is likely to accelerate in the current century. The scope and pace of such changes presents a severe challenge to societies and systems in the world. Society's adjustment to technological change often lags far behind, creating tensions and stresses. This also happens in the field of diplomacy and international relations.

The current century is also likely to witness unprecedented and rapid technological changes. Fundamental scientific research is likely to require increasingly greater resources, demanding greater efforts to convince non-scientific community of the benefits of such research, in order to secure support for funding. This may slow the growth of such projects, and indeed projects such as the Superconducting Super Collider and the ITER have already faced resource problems. While in the abstract, unraveling the mysteries of nature may excite the imagination of scientists, down to earth considerations of practical or strategic applications also play a role especially in securing support for research. For example, the increasing cost and time to achieve new discoveries in particle physics may indicate that we are approaching some kind of limit on human capacity to build large scale research facilities.

However, scientific research in areas that are less resource intensive and offer greater potential benefits to society is likely to accelerate. Examples of such areas are in biological sciences, nanotechnology, information technology, ocean science, energy, nuclear science and space science. Biotechnology and nanotechnology in particular offer vast prospects for new discoveries and practical applications with relatively lower investments. There is likely to be synergy between developments in nanotechnology, biotechnology and information technology which may open the way to dramatic developments in these fields, leading to far reaching changes. For example, nanotechnology may result in increasingly more powerful and compact informatics systems and lead to artificial intelligence and robotics applications not imagined in the past. Another possibility is the development of new materials with novel properties, new energy conversion and storage systems making it possible to exploit solar energy on a large scale and end our dependence of fossil fuels. In biotechnology, mastering the ability to create synthetic genomes and modify natural genomes may create a revolution in human health care with better treatment for cancer and degenerative diseases, organ and tissue replacement, and extending human life spans. There is vast scope for human ingenuity to make new discoveries in these fields. At the same time, there is the challenge of helping civil society and policy makers come to terms and understand the wider implications and consequences of technological development and take informed decisions.

## Diplomatic challenges

New technology often brings in its wake disruptions and social changes. These disruptions are transmitted far more rapidly across the globe in the current age of globalization. The challenge is to adjust and adapt to disruptions and make best use of the opportunities afforded by technological changes. There may also be a backlash against technology, driven by lack of public knowledge and fear and suspicion of the impact of new technology. Countries and businesses which are ahead in this game will improve their prospects while those which lag behind will lose out. Therefore, as technology development accelerates, it will become more and more important to plan ahead and adjust more rapidly. Governments will therefore need to pay greater attention to technology forecasting and early preparations to deal with likely changes.

New technology can offer leap frog opportunities to countries that were earlier left behind. For example, mobile telecommunications and the internet offer such a possibility with millions of low income users being able to exploit such services. New nanotechnology based production systems integrated with informatics can be more efficient than traditional manufacturing systems in future. Given the imperatives of climate change, developing countries may no longer be constrained to follow the traditional high carbon intensity paths to development and may be able to benefit from alternative paths to development.

Technology may amplify some of the existing challenges. Security and weapons proliferation including to non-state actors will remain important. Environment protection and climate change is another key area. Technology for development especially to achieve the Sustainable Development Goals 2030 will be another important issue. Regulation of technology to reconcile the interests of various stakeholders within and across countries will become increasingly complex. The challenge of applying science and technology for inclusive human development and reducing inequalities will need to be met

In the above scenario, science and technology are likely to play an increasing role in diplomacy and international relations[1]. An increasing array of scientific issues will demand the attention of diplomats and policy makers, especially as transboundary impacts are felt. This will require more scientific and technology awareness on the part of diplomats on the one hand, as well as closer coordination and cooperation between agencies involved in diplomacy and in science and technology development. Foreign ministries may need to devote more attention and intellectual resources to science and technology issues at an early stage, in order to exploit the diplomatic opportunities afforded and minimize potential losses. It will be necessary to have the optimum mix of scientists and or diplomats with scientific background and training posted in Embassies. Better communication between networks of diplomats and of scientific community would be helpful. It will also be necessary, in this age of rapid communications, to monitor and analyze opposition to introduction of new technology, as this can spread rapidly across the globe.

Could science and technology help in reducing differences among countries? This is similar to using nongovernmental actors, cultural diplomacy, and people to people contacts to bridge gaps or strengthen relations. At the basic level, it should be possible to do so, especially if science and technology can expand the pie of mutual benefits that can be shared. Scientists can often communicate better with their counterparts and help bridge communication gaps between estranged countries, depending on how much influence they wield in their respective countries. As science and technology advances, there may be more such opportunities for using science for diplomacy, and to help resolve long standing conflicts, as well as for stimulating greater cooperation among nations. This will require creative and imaginative interaction between diplomats and scientists in order to fashion useful and productive initiatives.

How can diplomatic machinery be better used to maximize the benefits of science and technology for societies? This is sometimes called diplomacy for science. The diplomatic network of a country could help in identifying promising new technologies that could be useful for the national interest. This requires closer interaction with the science, technology and innovation ecosystem, to spot possible opportunities and promising entities for developing useful linkages. In the case of India, the diaspora in advanced countries such as the US is a valuable human resource that can harnessed to stimulate interaction with the actors in the Indian S&T ecosystem for mutual benefit. Diplomatic machinery can also be used to barriers and controls to access to technology and equipment and mobility of researchers and students in technology areas of national interest. In the field of development cooperation, science and technology can help developing countries in achieving the sustainable development goals 2030, for example.

Science diplomacy will continue to evolve to keep pace with the rapidity and intensity of technological changes sweeping the globe and accelerating as we move through the current century. Diplomats, scientist and civil society have important roles to play in this evolutionary process.

# Endnotes

1  In 2010, the Royal Society and the American Association for the Advancement of Science (AAAS) had  identified three broad tracks in science diplomacy: (1) Science and diplomacy which involves scientific advice and inputs into diplomacy,  (2) science for diplomacy which involves using science to improve relations, and (3) diplomacy for science which involves using diplomacy to advance science

# Index

www.ingramcontent.com/pod-product-compliance
Lightning Source LLC
Chambersburg PA
CBHW020458270326
41926CB00008B/651